Advances in Asian Human-Environmental Research

The *Advances in Asian Human-Environmental Research* series aims at fostering the discussion on the complex relationships between physical landscapes, natural resources, and their modification by human land use in various environments of Asia. It is widely acknowledged that human-environmental interactions become increasingly important in Area Studies and development research, taking into account regional differences as well as bio-physical, socio-economic and cultural particularities. The book series seeks to explore theoretic and conceptual reflection on dynamic human-environment systems applying advanced methodology and innovative research perspectives. The main themes of the series cover urban and rural landscapes in Asia. Examples include topics such as land and forest degradation, vulnerability and mitigation strategies, natural hazards and risk management concepts, environmental change, impact studies and consequences for local communities. The relevant themes of the series are mainly focused on geographical research perspectives of Area Studies, however there is scope for interdisciplinary contributions from various spheres within the natural sciences, social sciences and humanities. Key themes: Human-Environment Interaction - Asian regional studies - Asian geography - Impact studies - Landscape - Society - Land use and land cover change - Natural resources *Submit a proposal*: Proposals for the series will be considered by the Series Editor and International Editorial Board. An initial author/editor questionnaire and instructions for authors can be obtained from the Publisher, Dr. Robert K. Doe (robert.doe@springer.com).

Huda Javaid

Land Commodification, Urbanisation and Social Change in Peri-Urban Villages, Lahore

Springer

Huda Javaid
South Asia Institute (SAI)
Heidelberg University
Heidelberg, Baden-Württemberg, Germany

ISSN 1879-7180 ISSN 1879-7199 (electronic)
Advances in Asian Human-Environmental Research
ISBN 978-3-032-10461-8 ISBN 978-3-032-10462-5 (eBook)
https://doi.org/10.1007/978-3-032-10462-5

This Springer imprint is published by the registered company Springer Nature Switzerland AG
The registered company address is: Gewerbestrasse 11, 6330 Cham, Switzerland

This book grows out of my doctoral work, 'Can Money Compensate?' at Heidelberg University, under the guidance of Prof. Dr. Marcus Nüsser.

To Dr. Munir Ghazanfar and Rabia Nadir—your mentorship and unwavering faith across more than a decade made this journey possible; these words are merely an echo of my gratitude.

Thanks also go to the Department of Development Studies, School of Social Sciences and Humanities (S3H), NUST, for the space and support that carried this work to completion.

Monologue

I walked into fields that are no longer fields.
The canals still whisper, but their banks are wounded with concrete.
Where wheat once swayed in the wind, walls now rise—tall, uniform, indifferent, striped like sentries guarding a world being unmade.
The earth that once fed families lies scraped bare, its topsoil sold, its fertility erased.
They call it development.
They measure it in rupees and square feet, in master plans drafted far from here—
as if a new partition is being etched again, cutting through the heart of community with an unfeeling hand.
But the people I sat with measured it differently:
in the silence of a courtyard where no children run anymore,in the grief of a farmer who sold his animals, his only living bank,
in laughter that now feels borrowed, stitched together to hide wounds,
in the absence of shoulders that once carried one another's burdens,
in a language slipping from the tongue, once an anchor of identity,
in the burial of an ecology that once felt like home,
in the fading echoes of girls' giggles that once lit up the fields,
in the disappearance of their sprinting steps, motherly earth smiles faintly,
bearing witness as even childhood is reclaimed by the fields.
in women whose bright eyes now dull behind the beautiful bars of modern houses that offer loneliness instead of liberation,
in graveyards where even the dead seem unsettled -
as if made to die twice: first in life, then in the quiet violence of development's erasure.
Cash came quickly, like a flood.
But it dried quicker still, leaving behind not wealth but emptiness.
You cannot buy back a world once it is taken.
You cannot put a price on a life-world once it is undone.
And yet - mid this unraveling - memories rise with a stubborn pulse.
Women remember the songs they sang during harvest.

Men remember the shade of mango orchards long gone.

Young people chase city dreams yet carry soil-stories in their pockets, the last fragments of a disappearing world.

Nostalgia lingers not as weakness,
but as quiet resistance -
a refusal to forget what developers so easily auction off.

This book is written from those fields, from those courtyards, from those memories.

It is written with my feet in the dust and my eyes on the ruins -
not only the ruins on the surface, but the ruins buried underneath,
beneath the silence, beneath what is left unsaid.
The accounts do not whisper; they scream.
They demand to be seen, to be heard, to be witnessed.
They challenge the very process that consumes them,
even as, from afar, they may appear like its silent accomplices.
This is not merely an account of what was lost.
It is a testament to what refuses to vanish.
Because even as concrete hardens,
the echoes of a breathing yet dying land,
of agrarian life,
of the painstakingly built asset called community,
still rise to ask us:
What is the worth of a world, and who pays its price?

Acknowledgements

Finally, the most awaited moment has arrived: writing an acknowledgement, which reminds me that I have reached the point where my book is complete. I have waited since the beginning to come this far, and it is more important than ever to acknowledge all the invisible hands that have been pushing me to complete this book. Reading other people's acknowledgment that began with the mention of parents used to sound cliché to me. Now, however, I know extremely well that the list of people to whom I owe my gratitude cannot begin without thanking my parents. Thank you for never clipping my wings, for never weighing me down because of my gender, and for protecting me from the social pressures that society so often places on women. I am deeply grateful for the trust you placed in me and for the unwavering support that allowed me to step into the world with confidence, knowing that nothing would be allowed to stand in my way.

Furqan Bhai, Nida Baji, and *Saad* and how can I not take your name, Iram Baji-this note is only a small token of appreciation for everything you have done. Your love, steady presence, and boundless encouragement are, and will always remain, far beyond anything I can adequately acknowledge. Together, you are the pillar upon which my life finds its balance, strength, and grounding.

I am immensely grateful to the DAAD for enabling me to conduct this study and for giving me the opportunity to open a final view of this land and community for everyone before it vanishes and becomes part of history. I would first like to thank my supervisor, Professor Dr. Marcus Nüsser. I am deeply grateful that he gave me the opportunity to work under his auspices and to be a part of his work group. He was kind throughout and generously supportive be it planning fieldworks, administrative issues, or academic tasks. I am glad and thankful; I got a chance to work with him. Here, Dr. Susanne Schmidt developed maps for me despite her highly busy schedule and multifarious responsibilities. She also taught me writing scientific papers through her many feedback, a skill for all time, that also helped in writing my book. The two also gave me a chance to teach at South Asia Institute (SAI), which was a great experience and helped me develop confidence for life. I am not sure how much I was able to impart but I gained and learnt a lot from the experience for which

I cannot thank you enough. I am also obliged to Professor Dr. Rahul Mukherji for coming on board as a second supervisor.

Furthermore, I would like to extend my appreciation for the MPhil programme at the Lahore School of Economics, especially Professor Dr. Munir Ghazanfar, for inculcating an attitude of critical thinking and learning from the people who experience the weight of our unjust and inequitable world. Many thanks to Dr. Nawaz Chaudhry, under whom I started my exploration of development in the Barki villages. To my decades-long friends who have stayed with me through all thick and thin—Fizza, Huma, Schiza, and Hina: I once read that the richest person is one who has one loyal friend, and I have many. I heartily thank you all for being there and for bearing with me.

Rabia Nadir was my local supervisor in Lahore. She taught me how to develop a rapport with the peasants and win their complete trust. Though I was an outsider in many ways, without her guidance, they certainly would not have opened their hearts to me. She personally accompanied me on many field trips and also guided me in data collection from offices, officials, and real estate dealers. For Rabia, it wasn't just her pupil; she was even more interested in the discoveries about the peasant society that my research was leading to.

This section would be unfinished without special regard for Mr. Nadeem, who did an incredible job of transcribing interviews for me. I am also grateful to Mr. Basharat and especially Mr. Tariq Gill, who unconditionally helped me collect empirical data through arranging meetings, introducing me to interlocutors to gather data, accompanying me on field visits every now and then by setting aside all his other duties and staying connected even after I left Lahore post-fieldwork. My driver, Mr. Afzaal, was very punctual, dutiful, and committed not only to his job but gradually to my research as well. He took me to places and chaperoned me on field trips. Gradually, he started finding people and persuading them to have long interviews with me. He joined me for many interviews and took an active part in the ongoing conversations at the time.

Before I conclude, I would like to share my impression that the help I received in Pakistan was not motivated by any material reward or advantage; in fact, it was not even an obligation. People were pleased to assist in a non-profit, pro-public activity.

Competing Interests The author has no competing interests to declare that are relevant to the content of this manuscript.

About the Book

Peri-urban villages on the fringes of Lahore are undergoing rapid socio-ecological transformation as agricultural landscapes give way to speculative real-estate development and expanding concrete frontiers. While similar processes are unfolding across the Global South, with parallels in parts of Latin America and Africa, a rich body of scholarship has examined these transformations through lenses, such as displacement, dispossession, urban inequality, informality, gentrification, land acquisition, infrastructure provision, environmental degradation, financialization, and related dynamics. Far less is known about how these transformations are lived and understood by village communities themselves as historically grounded and relational processes that unfold unevenly across space and time, particularly through changes in the intangible assets that shape quality of life.

This book brings forward the voices of those most affected yet least heard. Through the narratives of villagers navigating the intrusion of cash economies and infrastructure-led development, it traces the gradual unmaking of ecological, social, and cultural worlds that once anchored everyday life. These accounts complicate dominant assumptions about compensation and progress, revealing money as a pseudo-equalizer that obscures losses that cannot be priced, including land-based relations, trust, reciprocity, collective care, memory, and a sense of belonging rooted in shared ways of living.

Approaching peri-urban transformation through a political ecology lens, this book reworks Andre Gunder Frank's theory of the development of underdevelopment beyond its classical formulation of core–satellite relations between nations. It shows how the same structural logic unfolds along urban–rural frontiers, where expanding cities operate as cores and villages are rendered their internal satellites. Underdevelopment here is not produced through extraction alone, but through the reorganization of deprivation as land, ecology, and social life are systematically reconfigured in the name of development.

This book argues that underdevelopment is neither purely economic nor external. It is lived, relational, and cumulative. As cash economies penetrate peri-urban villages, communities experience material inflows alongside the steady depletion of assets that once sustained collective resilience, including common land, livestock, reciprocal labour, trust, care, and ecological security. What had existed in abundance outside the market is gradually commodified, fragmented, or lost. Even where households move into larger homes or upscale developments, many come to recognize a deeper dispossession, marked by declining well-being, weakened social bonds, and growing dependence on markets for survival.

Through deeply grounded ethnographic engagement, this work advances the concept of 'internal underdevelopment', showing how development can make formerly resilient communities underdeveloped from within. It challenges the dismissal of nostalgia as mere sentiment, reframing it instead as an epistemic register through which villagers articulate what quality of life once meant and what has been irreversibly displaced in the pursuit of progress. At once a scholarly intervention and an act of preservation, this book documents the final moments of a dynamic agrarian way of life on the verge of erasure, offering a critical rethinking of development, well-being, and what it means to live well.

Tagline

Preserving the last echoes of communities reshaped by capital, development, and concrete.

Contents

Chapter 1
What Is Lost When Land Becomes Cash?

Contents

Abstract This opening chapter offers a glimpse into the arguments and narratives that unfold across the book. It examines the socio-ecological unmaking of agrarian life-worlds in the peri-urban frontier of Lahore, Pakistan, where fertile farmland has been rapidly converted into elite housing schemes. Drawing on ethnographic research conducted between 2014 and 2022, it traces how cash for land appears as compensation and opportunity but simultaneously dissolves reciprocal ties, corrodes trust and undermines the moral economies that once sustained village life. The analysis foregrounds both visible disruptions as land acquisition, topsoil mining, ecological degradation, displacement and invisible fractures like the unravelling of kinship, the erosion of safety nets and the fading of collective memory. What emerges is not simply development, but a socio-ecological unmaking, where landscapes are commodified and social worlds rendered precarious. By situating Lahore's peri-urban expansion within global debates on neoliberal urbanism, forms of dispossession, land grab and political ecology, the chapter shows how development is experienced less as progress than as a political project of uneven ruination. Villagers' narratives, often dismissed as nostalgia, instead stand as testimonies of critique—records of loss, memory and alternative visions of justice that challenge the logics of commodification. Hence, serving as both an introduction and a concep-

H. Javaid, *Land Commodification, Urbanisation and Social Change in Peri-Urban Villages, Lahore*, Advances in Asian Human-Environmental Research, https://doi.org/10.1007/978-3-032-10462-5_1

tual map, it sets the stage for understanding how development, urban expansion and ecological change converge to reshape everyday life on the city's contested edges.

1.1 Visible and Invisible Change

The landscape of our village has transformed; it resembles a hilly area now (Javaid 2015).

It was reported by a villager and witnessed later by the researcher that gave a new direction to MPhil book which began in 2013 but also laid foundation for the later research. It was the beginning of ecological transformation in the village with removal of 1.5 m of topsoil from 0.405 ha of productive farmland. Removal of top soil was the most visible and manifest ecological disruption in the changes brought about by the shift in land use from agricultural production to elite suburban housing estate. It was the creation of a new landscape with attendant levelling of land, raising of streets, planting of new vegetation and laying of infrastructure of drain pipes, electrical transformers, cables, etc. This transformation was in line with neoliberal capitalist restructuring since the 1990s which had been driving commodification of rural outskirts of cities as land is acquired for development and investment under policies promoting economic growth through urbanisation (Myers 2021). This urban expansion has caused irreparable social and environmental deterioration through virtual land grab such as emergence of 'urban' diseases, food insecurity, amplified class differences, dispossession and development-induced displacement (Neef and Singer 2015; Zoomers 2011). Ecologies, human and nature vary from place to place, but human and natural monocultures are proliferating under an unchecked model of capitalist growth in Global South, Latin America and Africa as 'global recolonisation' (McMichael 2016). These processes have been analysed in great depth in multiple locales in the south. However, most studies especially in geography and environmental studies have focused on the ecological impacts but do not focus on the social aspects of this transformation especially as it is experienced and expressed by the communities directly affected by the new development (Abdullah 2023; Basheer and Waseem 2022; Chen 2024; Verma and Das 2024).

Pakistan has followed the same trajectory, and the periphery of metropolitan cities has assumed profound importance for investment for profit-making purposes. For example, land-use change for development in Karachi has thrown agrarian-pastoral *Baloch Goths* (villages) into disarray and induced huge displacement with support and facilitation by state policies and action (Anwar 2018; Hasan 2015). Real-estate development on the outskirts of Islamabad (Akhtar and Rashid 2021) and removal of mango orchards for the construction of gated community in Multan (Batool 2021) have similarly usurped huge swathes of rural land and removed the poorer inhabitants. Pakistan has been known for having the fastest paced urbanisation in South Asia with not only an annual rate of 3% but is expected to have half of the population in cities by 2025 (Kugelman 2013). Especially, in times of development in neoliberal epoch that treats villages as empty, blank canvases, for the city expansion. The rezoning of a city through city masterplans brings off-limit areas into available sites for investment and real-estate development (Goldman 2020). It was further verified by a direct interview with Lahore Development Authority (LDA) personnel involved in developing Lahore Masterplan 2050 (LDA, 2021). The city masterplan 2050 was all about the expansion of city that includes villages only as sites for infrastructure development but not as socio-ecological realms comprised of communities.

This study focuses on peri-urban villages southeast of Lahore where development of elite housing society took on rapid pace around 1999. After partition, Amritsar and Lahore got 84 villages, respectively. With time, several new villages were formed, but the 'area of 84 villages' denotes historical villages.[1] The villages on the Eastern periphery of Lahore city have been under the radar of the Defence Housing Authority (DHA). Presently, these villages are in varying stages of socio-ecological change as most land has been acquired through cash compensation, and development of elite gated housing is underway. This transfer of land gained momentum when Lahore Cantonment Cooperative Housing Society (LCCHS)[2] was revamped as DHA. This reorganisation of LCCHS happened amid a larger structural adjustment of economic liberalisation on the state level. A major policy plank of the neoliberal agenda was urbanisation and development of cities as engines of growth (Y. Huang 2022; Magdoff 2013; Zoomers 2010).

The research is an extended ethnography from 2014 to 2022 and offers an empirical record of perceptions, experiences and material changes in the life of villagers when land was taken in return for cash. The loss of land and new landscape is a material change, but there was also loss of intangibles as communities were disrupted by physical displacement and psychological reorientation as money became the most powerful arbitrator shaping space and social position. The loss of intangibles such as reciprocity, sharing resources both material and social, security, identity and cultural capital for the largely vulnerable, landless peasants, wage workers, elderly, females and children have not been recorded and theorised in the wide-ranging literature on socio-spatial transformation in South Asia.

In 2014, these villages offered a rich sampling of settlements at different stages of social and ecological transformation as one by one villages were being reformatted into a new urban space. At on one end of a spectrum of changing settlements were those villages which had been totally obliterated from the landscape, and on the other end were those waiting in anticipation for the buyers to arrive to purchase their lands. In the middle were villages which had lost all agricultural lands and were now walled ghettoes amid the new elite housing estate.

[1] The Settlement reports of UP and Punjab and other works discuss local village groups comprising of two to a theoretical of 'eighty-four' villages which had agnatic (paternal/connected by kinship or marriage) origin. There are number of names in vernacular terms such as *tappa, ilaqa, khera* and *mauza* (Marriott 1953).

[2] LCCHS—a cooperative housing society developed for the retired army officers by the Pakistan Army in 1975. Since then, it has become self-governing real-estate developer. Later in 1999, it was converted to DHA (https://dhalahore.org/about-us/history/).

The findings of this book can be framed by the epigram 'development of underdevelopment' (Frank 1966) as articulated by the victims of development in urbanising frontiers in the Global South (GS). The processes happening on the outskirts of metropolitan Lahore are part of the chain of accumulation in the core from peripheries in the Global North (GN). Local elites gain and lead this process, in this case through the development of suburbs for the rich by acquisition of cheap agricultural land from peri-urban villagers. The monetary gains and new landscape are experienced as underdevelopment of social well-being previously enjoyed by the majority, cash-poor communities in the peri-urban villages.[3] The research records the narratives of those who have experienced a rapid social change. Their narratives speak of a loss of intangibles and their degradation (both in terms of social and environmental from which there is no reversal) in this development which has brought tremendous material wealth to those who owned land. It argues that the cash received for land is incapable of buying community, values such as care, obligation, trust, brotherhood, security and love amongst a social group. Lahore has been expanding into its agrarian hinterland since the time of the British colonialists. However, ever since DHA appeared as an aggressive player developing private housing societies for the affluent (billboards installed in DHA announce 'aesthetically designed for community living' or 'DHA Lahore is a high-end living society'); the price on offer and the clout of the military have literally bulldozed a new landscape in the area (Ahmed 2023). The book records the villager's narrative of 'environmental and ecological change' in a frontier of speculative development. The villagers across class, gender, generation and ethnicity shared a compelling assessment of socio-ecological decline. They tell a story of loss and the role of money and power of state (i.e. the role of military in aggressive support of urbanising processes) in the making of a new environment and the marginalisation of the weakest. It contends the easy dismissal of 'nostalgia' for the lost village and 'land' and iterates the importance of the 'voices' of the most affected. The communities in these villages and ecologies are presently trapped in the capitalist growth model, going through unfathomable change often dismissed by labelling them as nostalgia.

A significant body of literature addresses multiple aspects of this unbridled urban expansion: displacement and dispossession experienced through the emplacement (Anwar 2012; Neef and Singer 2015; Tassadiq 2022), the role of state and its power (Akhtar 2006; Dwyer 2022; Gazdar 2009; Hussain 1984; Levien 2017; Levien and Upadhyay 2022) or how the change in socio-economic patterns has modified the use and outlook towards land through rural urbanisation in southern Punjab (Mughal 2019). This study records and analyses the invisible social change as perceived by the locals and juxtaposes it with more evident material change. Given that society

[3] Big houses, open land, communal land includes wastewater ponds, shrines, graveyards, marriage houses, etc. The availability of clean drinking water, good quality of food, space to keep animals as source of dairy and the access to abundant free fodder for animals to keep. The safety and sense of security coming from a community that was built over time, a relation of barter or payment in form of services and kind, due to which the subsistence was intact and people were not fully at the mercy of market and cash economy.

and nature or human and environment are intertwined, hence, commodification of land is also transformative of social relations which changed as monetary exchange value recalibrated the worth of everything that provided sustenance, shelter and security to indigenous communities. An array of resources of use-value such as land-based produce accessed through traditional social arrangements and multi-use commons acquired monetary value and became inaccessible to communities causing disruption of existing social and ecological formations.

Peri-urban villages are microcosms subject to rapid transformation brought by massive cash influx entering rural communities as a monetary compensation for the land. The transaction however entails not just loss of land but also the agrarian life attached to it. The compensatory packages cannot compensate for the ecology which is lost due to profit-driven development. The periphery and communities are the living, breathing proof of urban expansion as a failed development policy that has jeopardised natural resources such as fertile soil, productive land and water and sabotaged the social realm by fragmenting communities through cash induction. The commodification of land has in short destroyed relatively stable socio-ecological zones, and cash cannot compensate for this on-going underdevelopment.

1.2 Southeast, Lahore as a Research Site—Why?

Lahore is the capital city of Punjab province, located in the northwest of Pakistan with an estimated population of over 14.4 million in 2024 with annual change of 3.06% (Pakistan Bureau of Statistics 2024). In the past five decades, the built-up area in Lahore has expanded to 532 km^2, whereas this urban expansion has happened at the expanse of fertile agricultural lands, vegetation and other cultivable lands (Minallah et al. 2021), which is expected to be 1320 km^2 by 2025 (Ibrahim and Riaz 2018). DHA alone is rapidly expanding towards southeast and east, covering almost 20,000 acres across its nine phases in the given direction (Abbas and Wakil 2023). It was once the restricted area for the development in order to save agricultural lands and to protect the border.

Punjab as a region has undergone massive transformation since Britishers took over (Talbot 2011) through land privatisation, transformed agriculture, enclosure of commons and extraction of resources (Ranauta 2020). Especially the Canal Colony projects unbridled the process of urbanisation in Punjab through agricultural colonisation (Singh and Singh 2019). Historical texts show that suburban sprawl into rich agricultural lands in Lahore is not a new process, as its roots go back into mid-nineteenth and first half of twentieth centuries when the colonialists started development of the city for their new political and economic needs. Lahore witnessed the developments of civil lines through residential quarters for the colonial bureaucracy, cantonment, railway infrastructure and colonies and modern residential areas mainly for middle and upper class consumption (Chattha 2012).

This study focused on an area which had seen a very slow and limited urbanisation, villages in close proximity to the city centre and well connected to the central city through major roads. The area has seen large-scale land transformation in a relatively short span of time from agricultural land to suburban developments and particularly of the most elite social class in the country. Most of these villages situated in the area, are historical in nature, existed prior to partition are the carriers of tales of partition. Many of these villages were named after *Sikhs* and are still known by the same names such as *Sarja Marja, Natha Singh, Mota Singh, Heera Singh,* etc.

Typically, these villages were hubs of land-based production and considered some of the most fertile lands in the Punjab (Zaman and Baloch 2011). This production consisted of not only major crops such as wheat, rice, corn and fodder but also vegetables such as carrot, pea, radish, green legumes and potato and fruits such as guava, strawberries, melons, etc. However, almost half the population of these villages did not own land and had diverse livelihoods mostly linked to agriculture and livestock. In addition, they did wage work in Lahore city and carried on petty production and small-scale commercial enterprises in their spacious homes. Cash was generated by the landless and small peasants through produce which was sold in the cities including milk, chickens and cow dung cakes. They were able to supplement their cash incomes with in-kind acquisition of grain through harvest labour, free firewood and fodder from commons such as graveyards and greens along irrigation and drainage channels. Most significant to their subsistence was the social capital which allowed them to secure grain on loan from farmers. This subsistence remained secured till the DHA arrived in the area and started taking agricultural lands. This land acquisition spurred by the DHA after the building of new terminal of the Lahore's international airport. The community of settlements had created a larger set of relationships amongst each other, with land they lived on, the livestock which they looked at insurance and the source of basic dietary items. These villages can be described as suburbs of the poor wage workers, entrepreneurs and petty producers which provided a rich and secure support environment for children and females.

The trend of elite suburban development with its extensive road network and walled enclosures shifted around 2002 from being mostly towards south of Lahore to the south-eastern periphery, the site of this study. The presence of the cantonment and proximity of the border with India had till then constrained development and restrained the creation of a market in land for housing in these villages. This restrain (proximity to the Indian border and needs for approvals by the military) on growth on the eastern periphery of Lahore ended once the army itself became a major land development agent in the area.

Initiated in 1977, the first four phases of this housing with the total area of 11.3 km^2 were developed from 1978 to 1993 (Zaman and Baloch 2011).[4] Where first

[4] Phase 1, commenced and completed during 1978–1982. Phase 2 (1981–1984), Phase 3 (1983–1988), Phase 4 (1986–1993), Phase 5 (2001), land acquisition was done between 2002 and 2003 for phases 6–10; however, phase 6 was officially launched in March 2005; phase 7 was launched in 2008, which is further divided into two phases: phase 1 of it was developed between 2008 and 2015, and the development of its second phase began in 2015 till to date. Whereas the built-up began around 2006–2007, DHA got the NOC approval for EME (Phase 12) in 2006. DHA acquired EME in 2003, massive development happened in 2012 and was named as DHA Phase 12; reports reveal that EME officially began in 2006. DHA Phase 8 development commenced since

Fig. 1.1 Examples of house prices in posh locality 'DHA': 0.05 ha house with full basement, DHA, 108.5 million PKR (360,406 €, top left), 0.05 ha house without basement, DHA, 68.5 million PKR (2,275,376 €, top right), 0.03 ha house, DHA, 50 million (166,086 €, bottom left), 0.01 ha house, DHA, 16 million PKR (53,147 €, bottom right). All pictures have been taken from https://www.zameen.com/Homes/Lahore_DHA_Defence-9-1.html. Prices of the houses vary based on phase, location, design and architecture

four phases took 15 years to be built, the remaining nine phases were developed in 24 years. The rapid pace of expansion of the elite housing on the southeast makes its distinctive.

The research focuses on expansion since 1999 when Lahore Cantonment Co-Operative Housing Society (LCCHS) was converted to DHA Lahore through a provincial ordinance. Later in 2002, it was federalised—taken as the birth of DHA Lahore (Ahmed 2023; DHA website n.d.; Siddiqa 2007). Since then, the southeast periphery of Lahore became a hub for business in land for DHA which opened floodgates for several other private developers such as Imperial Garden and Paragon (on the other side of *Barki* Road). Nonetheless, DHA has been considered as a brand in housing which is meant to meet the needs of the rich. It is a posh residential area and has been known as a status symbol. People living on a piece of land, called DHA, are sitting over millions of worth of properties (see Fig. 1.1); therefore, it is fairly easy to place them in the top social tier.

2016. IVY Green was launched in March 2015, and (sector Z) development started from July 2016, Phase 9 called Prism was launched in February 2016; DHA Rahbar (Phase 11), previously known as Rangers Society, was later on taken over by DHA and was launched in 2015, and phase 13 (DHA City) was launched in July 2010. DHA took 15 years to develop first four phases; however, the remaining phases and extension, i.e. 9 phases, were developed within less than 24 years. DHA Phase 10 has been launched recently, in April 2024. The information has been collected from DHA homepage, different websites and YouTube.

1.3 Contribution

Urbanisation brought transformation in the peripheral villages of Lahore over the long haul of a couple of centuries. Since the 2000s, it has happened rapidly in villages of southeast of Lahore. It also underscores that development can't be seen and followed by blind optimism that has legitimised resource confiscation from the GS, and now same can be witnessed from city-periphery dichotomy, the development as 'executioner' of villages, agrarian economy and rural ecology to facilitate the rich and wealthy. Yes, this is not a first study on development critique, yet the vast literature cannot pay much attention to bring voices of the most unheard to the fore ethnographically for the provision of empirical record of localised communities' perceptions and experiences.

Existing literature highlights physical, symbolic displacement and place attachment (Atkinson 2015; Bose 2020; Celestina 2016), the role of state and the violence of laws (Akhtar 2006; Barman and Chowdhury 2024), spatial transformation using remote sensing showing land use change/vegetation cover/surface temperature (Abdullah 2023; Zafar et al. 2024), groundwater as a resource and its extraction (Birkinshaw 2022), etc. A glaring omission is a record and theorisation of the narrative of the local communities—their everyday encounters with the process and its role in changing individuals and community, by restructuring and revamping space and place. Land is not just a physical landscape, but it is intrinsically social in character. Villages are the holders of unique ecology because there is no isolation and individualism in rural life. Land binds people, and people are obliged to mother earth that provides them livelihood, food and subsistence. Villagers in turn do not treat land as a commodity but feel responsible to take care of it. This brings people together because agriculture needs a manpower and labourer, and even in hierarchal structure, people from top to bottom tier are co-dependent. Agricultural labour and artisanal castes used to provide services to farmers, and landlords used to pay back for services. The villagers either learnt or inculcated the art of sharing and reciprocity which provided them safety net. People did not know physical and emotional hunger (discussed in detail in Chap. 5). The graveyards became a key contentious issue when realtors set their sights on burial grounds of emotional significance, the preservation of history, a sense of belonging to lineage and a secured piece of land after death. The world is crumbling, ecology is disappearing and communities are already starving for the sense of connectedness they had due to their relations. The quality of life is declining due to the loss of intangibles which they cherished the most. The intangible wealth can't be seen through naked eye, and the quality of life and how it is impacting cannot be judged by looking at physical manifestation of development through land-use change. But, in order to understand what development has brought to the table and what these marginalised communities have gained or lost, one must have to go deeper, put façade aside and develop trust and compassion for people to fathom the reality of development, to identify the real beneficiaries and the ones who are paying for it. The question of 'how' is of paramount importance and can only be addressed through ethnographic research which itself is a contribution and fills a methodological gap. Even today, geographers don't place ethnography as a method in the field, whereas the primary method of choice has been ethnography to study people their aspirations, fear, the power of resistance, opinions about the process across class, gender, generation and ethnic groups. Through this study, I also put emphasis on including methodologies that can help in finding answers instead of focusing on a particular field. It is an interdisciplinary project, and research should be beyond labels and tags. I am writing this book on behalf of all the knowledge producers (the interlocutors) who wore heart on their sleeves and allowed me to share part of their lives.

The other aspect that this work foregrounds is the rejection and dismissal of nostalgia. It is a powerful emotion that holds significant unheard stories of change and experiences which also provide a reference point to draw comparison with the present. There are hardly any studies available on sociology of nostalgia from south Asia, which deserves due recognition in sociology (Jacobsen 2023). I argue that nostalgia must not be seen superficially, and it should be given a due recognition in research that revolves around communities and their relation with development. It is easy to sweep it under the rug by calling it 'a memory of old good days', but it indicates detachment and lack of empathy for those who have been deprived of their traditional lives. Nostalgia itself is the documentation of change that is written on every individual's mental landscape and is also an indication of active minds, very much alive and registering observations and giving eternity to people, place, relations and environment through all the five senses. The memories were once their real-life events and emotions. Therefore, nostalgia is nothing but the reality of their lives, that is why it provides a realisation of loss when they can no longer experience in present times. The communities find themselves caught in a tide of change - aware of the loss yet powerless to halt it. Physical relocation is not a prerequisite for such grief; emotional displacement caused by ecological transformation, captured in Albrecht et al.'s (2007) notion of solastalgia, can be just as profound. As the rural landscape recedes, people grow nostalgic and recount how these changes have unfolded beyond their control or comprehension. Perhaps it is no exaggeration to say that such nostalgia often gives birth to poets, folk stories, and new forms of poetry. A few Punjabi verses from a long poem about change, written by a local poet, are translated. The following excerpt is about a new face of a village post 2002:

> *Now my village is plagued by a new disease*
> *A businessman arrived desiring to buy land for money*
> *Roads and houses constructed, have surrounded the village from all sides*
> *Eyes no longer see the agricultural lands, gone are water channels and water wells*
> *mother earth which fed us our daily bread and butter is ruined*
> *All the gardens and orchards are destroyed and crop field are seen no more*
> *Everyone is chanting 'money, money', but no one any longer calls and remembers the 'poet'*
>
> (The 'poet' is replaced with the original name of the poet by a researcher to keep his anonymity.)

The change has also been recorded and made visible through the use of Remote Sensing and Geospatial Information System (GIS). It is an excellent tool to develop maps for the virtual representation of change in pixels. There were virtually no maps of Lahore's[5] villages, except for an old Russian-script map of the Lahore–Amritsar region, which was acquired from the Heidelberg University Library. I could not get any updated, recent maps dedicated to villages, which shows the neglect even on the level of cartography of village geographies. Lahore city masterplans do not show any villages in the peripheries that were going to be lost in the city expansion. Therefore, through this work, effort is made to develop maps to keep a history of these locales which has not been documented in recent times[6] and is in the process of being totally transformed. Remote sensing data need to be validated on the ground which otherwise would leave discrepancies in virtual and on-field data. My intimate knowledge and photographic documentation helped to create a reliable record of the terrain.

The sum and substance of the significance is that it is tale of environmental change echoing from the community of villages in a frontier of speculative development based on the narratives of the villagers. It is a story of melancholy for the majority and happiness for a few. It attempts to evaluate a local project of 'development' in light of the narratives and manifest change in the lives of villagers and perceptions. For small landowners, it did bring money but took away many intangibles that can never be commodified. Greater prosperity came to those who are already prosperous, but even their narrative of social change did not paint an entirely positive view of development. They however did appreciate the monetary gain and the advanced, modern amenities they had acquired. Other than economic class, the village communities are an amalgam of gendered, ethnically and religiously differentiated individuals. They in case of the villages are more directly attached to land-based production and deeply connected as community. The historical Punjab village was self-sufficiently functioning as a whole with its own production and local

[5] The one-on-one conversation with GIS expert in Lahore (2024) also confirmed the lack of work on mapping which is unfortunate.

[6] Before British came, India had several kingdoms as per political division, but village life tied with culture shaped 'India'. The Battle of Plassey in 1757 ended the Mughal rule and started the British rule in India, which was also the harbinger of surveying and mapping that continued well for over 100 years. The geography of India was not much known upto the eighteenth century, but several maps of India came out from France, Venice, England and Holland, based on information collected from people (e.g. traders, mariners, travelers, route surveys). The British came to Indian Subcontinent as it was rich in resources and biodiversity which attracted Europeans and East India Company to land in 1600 for trade purposes. Hence after, they established three presidencies: Bombay, Madras and Calcutta during the early decades of the seventeenth century for trade. Later two events (1757 battle and Tipu Sultan's demise in 1799) led to other two events in the history of geographical survey in India—the setup of Survey Department and the invention of Great trigonometrical survey (GTS) in India. Three major surveys—GTS, topographical and revenue surveys—were carried out in most parts of India during the nineteenth century that were integral for mapping from the field and village to country level. During the late ninettenth and early twentieth centuries, most of the villages (6.4 lakhs) were surveyed, and cadastral maps were prepared. These practices were closely linked to attempts of the British Raj to conquest, rule, control and administer resource exploitation. The information is taken from Kalpagam (1995); Meenakshi (2023) and Paudel et al. (2021).

services.[7] The encounters are influenced by the geographical location too, for instance, speaker from a ghetto, participants from walled off localities, those who live in a village and then the ones who have migrated (in situ or moving out to an upscale locality). In addition to that, *who is facing it and how* cannot be conceived unless the gender and age (young, middle or senior) with respect to their class are taken into consideration. They all come together to tell from first-hand experience to let others know the facts. They tell a story of loss and the role of money and power of state (i.e. the role of military in aggressive support of urbanising processes) in the making of a new environment and the marginalisation of the weakest. They lamented the loss of communities and a feeling of being displaced without physically displacement. They are nostalgic over the loss of *Mahoal*—an environment, they will never be able to recreate with one-time compensation money, because everything is not for sale! The nostalgia is real, an amputee will of course keep missing a limb which was once a part of his life. The loss of land, culture, traditions and community is an invisible amputation, and people recall it every day.

The work aims to bring into view evident material changes as well as the psychological and emotional toll that development processes take on communities and distort the social landscape in agrarian societies once tied to land and livestock. The planners and influential realtors are thriving on commodifying intangible assets out of which security is the most valued. Fear of the outsiders—new expenses and loss of identity—is now pervasive. The periphery and communities are the living, breathing proof of urban expansion as a failed development policy that has jeopardised natural resources such as fertile soil, productive land and water and has destroyed relatively stable ecological zones. The militarised colonisation of land has sabotaged the social realm by fragmenting communities where cash induction has played a vital role in deepening class cleavage, mutating social ties and inculcating sense of individuality supplanted by collectivism.

In order to understand and to ensure social justice, planners and state have to understand development from the perspective of those who suffer it. To critically grasp these processes, it is essential to let go of the notion that 'this is not new; it happens everywhere', ecologies are diverse and unique in nature and if something

---

[7] The agricultural land is immobile that binds communities to it and provides stability. It is life-bearing in plethora of ways. Land is the major means of production and subsistence. People even in villages in contemporary times are still not fully dependent on market economy, and their survival is dependent and managed within the village. Even the landless does not go hungry. The exchange was more in kind rather through money. A focused group discussion (2021) at *Pathanke's Lambardar* house, a *Patwari* explained, *'People in villages are more dependent on animals than the produce as yield happens in every 6 months whereas, through animals, one can earn on daily basis. But the thing is that the debts are paid through services in kind. For instance, in Padhana, loan is also in form of serving a cup of tea for 6-month (time between from one produce to another) let us say. It can be done through sugar as well. We say, when I will get the wheat harvest, you may take one or two sacks of it, in exchange serve me tea for a year. This is how artisanal castes take wheat or fodder as payment for their services'*. Money is not involved in villages as it is in the outside; therefore, their reliance is more on community based on trust, keeping word, courtesy and resources coming from an agrarian setting. Therefore, there is degree of stability in opposition to what cash economy offers, i.e. volatility and uncertainty.

happens everywhere—it gains a lot more significance and screams for not just serious but sincerest attention, a need for recognition and acknowledgment instead of getting used to it. Hence, this study investigates how peri-urban communities experience urbanisation as a state-led development project, where the marketisation of land and the arrival of compensation packages generate profound socio-ecological shifts. It traces not only the visible transformations but also the changes in deeper, subtle, yet powerful intangible assets that, though unseen, anchor and give meaning to the agrarian way of life.

1.4 Research Question and Objectives

Can cash compensate for loss of community and ecological destruction in peri-urban villages of Lahore?

The objectives are:

- To determine the role of cash as compensatory compensation in restructuring space and society in peri-urban villages in Lahore
- Document change in socio-ecological worlds based on local voices, especially females in rural communities
- Situate the development of elite housing in peri-urban villages within larger processes of urbanisation on national and global levels

1.5 Structure of the Book

Chapter 1 introduces the study, its significance, reason for choosing it, central thesis, objectives and how the thesis has been structured.

Chapter 2 gives an overview to the study area where research has been conducted. This chapter also describes the history of urbanisation in Lahore. However, this chapter also compares spatial and temporal change happening due to driving forces such as economic developments and presence of power structures and also sees how it is different in contemporary times.

Chapter 3 is about the theoretical framework in which the work has been theorised. It also highlights the relevant themes that have been taken into consideration as they deem relevant and quintessential to strengthen the argument. Also, this chapter fathoms urbanisation as a multifaceted process that has transcended the conventional definition, stating migration from rural to urban areas. Now, cities have expanded viciously in pursuit of a uni-directional Western model of development, recklessly ignoring and dismantling the uniqueness and local ecology embedded in environment and culture. Power and greed are breaking and disrespecting the boundaries of stable ecological zones that once sustained them, and which are now bearing the brunt of hollow developments and are helpless in facilitating the myopic view of economic growth. This view sustains the lie that neoliberal economic policies implemented by the State want people to buy and believe, regardless of their will.

Chapter 4 is a detailed description of methodology adopted to carry out research. It is a mix of qualitative methods—however, ethnography was the primary method of gathering empirical data on social change. The physical locale was mapped to show the change in land historically and in terms of ecological aspects that have mutated and quantitative. There is a dire need for mapping as maps are not available in any significant detail or variety for the city and particularly on rural areas in Pakistan. The purpose of making maps for the virtual representation of change over the long haul is an attempt to give due importance to villages with their proper placement on map. It will be a contribution to cartography. Chapter 4 provides a detailed account of ethnography, the purpose of choosing it as a primary method and why it turned out to be a tool to fix all the pieces of the jigsaw puzzle. The reservations due to researcher's positionality and limitations have also been exposed in both quantitative and qualitative research methods.

Chapter 5 is the soul of book as it illustrates results being drawn from the data collected from the people in their natural, lived environments. The chapter has two distinctive halves such as physical manifestation of land-use change and the social degradation and how both these paramount features are deeply connected to form an ecology. Both physical and social transformations are further classified into sub-themes. Results are supported by using quotations from interviews, and arguments are referenced by the ones who were on-ground, real-life knowledge producers with face and voice.

Chapter 6 is the discussion chapter with the purpose of looking at similarities, but also the dissimilarities, by comparing it to other case studies available in the literature. It also underscores the area or aspect of the process that has taken different forms, shapes and names over time yet still not commonly talked about. It is significant not because there are few studies, but because very few examine these processes through people's own experiences - their everyday routines, aspirations, apprehensions, and whether these are changing or evolving with the influx of money. It also matters to understand whether the perceived gains and losses show any consistency, even when success stories exist. Cash, despite being framed as a symbol of power, may address certain immediate needs. Yet it cannot compensate for the forms of loss that undermine people's true sources of strength - resilience, resistance, and a sense of certainty - none of which the market world can provide, and whose inherent volatility only amplifies the insecurity people experience. Research also argues that such developments- through infrastructure building and the expansion of elite housing, which treat agrarian lands and communities as resource enclaves (Le Billon, 2001) - are in fact taking developing countries backwards and need to be reconceptualised. They are deteriorating the quality of resources and degrading lives by widening class cleavages, deepening disparities, and creating fractures within communities through the promotion of capitalist behaviours such as individualism over collectivism, and competition in place of reciprocity.

Chapter 7 is the last and final chapter of the book to conclude the study.

References

Abbas M, Wakil I (2023) Exploration and analysis of the drivers of urban sprawl in Pakistan: a case study of Lahore—RASTA. RASTA 66. https://rasta.pide.org.pk/cgp/exploration-and-analysis-of-the-drivers-of-urban-sprawl-in-pakistan-a-case-study-of-lahore/

Abdullah M (2023) Environmental impacts of urban sprawling and shrinking green cover in Lahore, Pakistan [fi=Ylempi AMK-opinnäytetyö|sv=Högre YH-examensarbete|en=Master's thesis|]. http://www.theseus.fi/handle/10024/812130

Ahmed A (2023) The rise of military capital in Pakistan: military neoliberalism, authoritarianism and urbanization. Geoforum 146:103846. https://doi.org/10.1016/j.geoforum.2023.103846

Akhtar AS (2006) The state as landlord in Pakistani Punjab: peasant struggles on the Okara military farms. J Peasant Stud 33(3):479–501. https://doi.org/10.1080/03066150601063058

Akhtar AS, Rashid A (2021) Dispossession and the militarised developer state: financialisation and class power on the agrarian–urban frontier of Islamabad, Pakistan. Third World Q 42(8):1866–1884. https://doi.org/10.1080/01436597.2021.1939004

Albrecht G, Sartore G-M, Connor L, Higginbotham N, Freeman S, Kelly B, Stain H, Tonna A, Pollard G (2007) Solastalgia: the distress caused by environmental change. Aust Psychiatr. Bulletin of Royal Australian and New Zealand College of Psychiatrists 15(Suppl 1):S95–S98. https://doi.org/10.1080/10398560701701288

Anwar N (2012) State power, civic participation and the urban frontier: the politics of the commons in Karachi. Antipode 44(3):601–620. https://doi.org/10.1111/j.1467-8330.2011.00920.x

Anwar N (2018) Receding rurality, booming periphery: value struggles in Karachi's agrarian-urban frontier. Econ Polit Wkly 53(12):46–54. https://ir.iba.edu.pk/faculty-research-articles/196

Atkinson R (2015) Losing one's place: narratives of neighbourhood change, market injustice and symbolic displacement. Hous Theory Soc 32(4):373–388. https://doi.org/10.1080/14036096.2015.1053980

Barman D, Chowdhury S (2024) Land for urbanization: shifting policies and variegated accumulation strategies in a fast-growing city in eastern India. Land Use Policy 140:107111. https://doi.org/10.1016/j.landusepol.2024.107111

Basheer MA, Waseem MA (2022) Spatiotemporal analysis of urban growth and land surface temperature: a case study of Lahore, Pakistan. The International Archives of the Photogrammetry, Remote Sensing and Spatial Information Sciences, XLVIII-4-W5–2022, 25–29. ISPRS TC IV: 7th International Conference on Smart Data and Smart Cities (SDSC)—19–21 October 2022, Sydney, Australia. https://doi.org/10.5194/isprs-archives-XLVIII-4-W5-2022-25-2022

Batool Z (2021) Horrible! Why DHA has cut about 6,000 acres of mango orchards in Multan? Parhlo. https://www.parhlo.com/6000-acres-of-mango-orchards-cut-by-dha-in-multan/

Birkinshaw M (2022) Grabbing groundwater: capture, extraction and the material politics of a fugitive resource. Geoforum 136:32–45. https://doi.org/10.1016/j.geoforum.2022.07.013

Bose S (2020) Attachment to place and territoriality in Nandigram land struggle, India. Hum Geogr 13(2):139–153. https://doi.org/10.1177/1942778620944383

Celestina M (2016) "Displacement" before displacement: time, place and the case of rural Urabá. J Lat Am Stud 48(2):367–390. https://doi.org/10.1017/S0022216X15001212

Chattha I (2012) Economic change and community relations in Lahore before partition. J Punjab Stud 19:193–213

Chen M (2024) Impacts of urbanization and climate change on ecosystems in Asia: challenges and conservation strategies. In: Third international conference on biological engineering and medical science (ICBioMed2023), 12924, pp 1063–1067. https://doi.org/10.1117/12.3021827

DHA website (n.d.) Histroy of DHA Lahore. DHA Lahore. Retrieved May 29, 2023, from https://dhalahore.org/about-us/history/

Dwyer MB (2022) Upland geopolitics: postwar Laos and the global land rush. University of Washington Press. https://doi.org/10.6069/9780295750507

Frank AG (1966) The development of underdevelopment. Mon Rev:17–31. https://doi.org/10.14452/MR-018-04-1966-08_3

Gazdar H (2009) The fourth round, and why they fight on: An essay on the history of land and reform in Pakistan. PANOS South Asia, Collective for Social Science Research, Karachi.
Goldman M (2020) Dispossession by financialization: the end(s) of rurality in the making of a speculative land market. J Peasant Stud 47(6):1251–1277. https://doi.org/10.1080/03066150.2020.1802720
Hasan A (2015) Value extraction from land and real estate in Karachi (pp. 181–198). https://doi.org/10.46692/9781447313496.011
Huang Y (2022) Neoliberalism in Asia: under the lens of land Acquisition for Development in India and China. Strateg Anal 46(4):441–458. https://doi.org/10.1080/09700161.2022.2111766
Hussain A (1984) Land reforms in Pakistan: a reconsideration. Bull Concerned Asian Sch 16(1):46–52. https://doi.org/10.1080/14672715.1984.10409782
Ibrahim M, Riaz S (2018) The urban Gazette: Lahore growth—past and future. Urban Unit. https://www.urbanunit.gov.pk/Download/publications/Files/9/2021/Urban%20Gazette%20Issue%202018.pdf
Jacobsen MH (2023) The sociology of nostalgia. Curr Opin Psychol 50:101556. https://doi.org/10.1016/j.copsyc.2023.101556
Javaid H (2015) Envrionmental impact assessment of Shuhada Town, DHA PHASE IX. [Thesis submitted to Lahore School of Economics in partial fulfillment of the requirements for the degree of MPhil Environmental Sciences]. Lahore School of Economics (LSE)
Kalpagam U (1995) Cartography in Colonial India. 30 (30), 87–98. http://www.jstor.org/stable/4403049
Kugelman M (2013) Urbanisation in Pakistan: causes and consequences. NOREF Expert Analysis Report
Le Billon P (2001) The political ecology of war: Natural resources and armed conflicts. Political Geography 20(5):561–584. https://doi.org/10.1016/S0962-6298(01)00015-4
Levien M (2017) From primitive accumulation to regimes of dispossession. Oxford Scholarship Online. https://www.academia.edu/12684353/From_Primitive_Accumulation_to_Regimes_of_Dispossession_Six_Theses_on_Indias_Land_Question
Levien M, Upadhyay S (2022) Toward a political sociology of dispossession: explaining opposition to capital projects in India. Polit Soc 50(2):279–310. https://doi.org/10.1177/00323292211016587
Magdoff F (2013) Twenty-first-century land grabs. Mon Rev 65(6):1. https://doi.org/10.14452/MR-065-06-2013-10_1
Marriott M (1953) Village structure and the Punjab Government: a restatement. Am Anthropol 55(1):137–143. https://doi.org/10.1525/aa.1953.55.1.02a00300
McMichael P (2016) Development and social change: a global perspective, 6th edn. SAGE Publications, Inc
Meenakshi CS (2023) The history of geographical surveys in India during the British period. Indian J Hist Sci 58(2):151–156. https://doi.org/10.1007/s43539-023-00088-y
Minallah M, Zia S, Rahman A-U, Riaz O (2021) Spatio-temporal analysis of urban expansion and future growth patterns of Lahore, Pakistan. Geogr Environ Sustain 14:41–53. https://doi.org/10.24057/2071-9388-2020-215
Mughal MAZ (2019) Rural urbanization, land, and agriculture in Pakistan. Asian Geogr 36(1):81–91. https://doi.org/10.1080/10225706.2018.1476255
Myers G (2021) Urbanisation in the global south. In: Shackleton CM, Cilliers SS, Davoren E, du Toit MJ (eds) Urban ecology in the global south. Springer International Publishing, pp 27–49. https://doi.org/10.1007/978-3-030-67650-6_2
Neef A, Singer J (2015) Development-induced displacement in Asia: conflicts, risks, and resilience. Dev Pract 25(5):601–611. https://doi.org/10.1080/09614524.2015.1052374
Pakistan Bureau of Statistics (2024) Lahore population 2024, https://worldpopulationreview.com/world-cities/lahore-population

Paudel B, Sapkota P, Pahari S, Pathak H (2021) Cadastral system in India. ResearchGate. https://www.researchgate.net/publication/354331381_CADASTRAL_SYSTEM_IN_INDIA
Ranauta J (2020) Transnational modernity/coloniality: linking Punjab's canal colonies, migration, and settler colonialism for critical solidarities in Canada. Stud Soc Justice 14(2):Article 2. https://doi.org/10.26522/ssj.v14i2.2272
Siddiqa A (2007) Military Inc: inside Pakistan's Military Economy. https://readersend.com/product/military-inc-inside-pakistans-military-economy/
Singh B, Singh B (2019) Punjab under the British rule: historicising the local transformations. Indian Hist Rev 46(2):207–226. https://doi.org/10.1177/0376983619889520
Talbot I (2011) The Punjab under colonialism: Order and transformation in British India. Journal of Punjab Studies 14(1):4.
Tassadiq F (2022) Producing dispossessed and humanitarian subjects: land acquisition and compensation policies in Lahore, Pakistan. PoLAR: Political and Legal Anthropology Review, n/a(n/a). https://doi.org/10.1111/plar.12506
Verma A, Das P (2024) Urbanization in the global south: perspectives and challenges. Vikalpa 02560909241234200. https://doi.org/10.1177/02560909241234200
Zafar Z, Zubair M, Zha Y, Fahd S, Ahmad Nadeem A (2024) Performance assessment of machine learning algorithms for mapping of land use/land cover using remote sensing data. Egypt J Remote Sens Space Sci 27(2):216–226. https://doi.org/10.1016/j.ejrs.2024.03.003
Zaman K, Baloch AA (2011) Urbanization of arable land in Lahore City in Pakistan: a case-study. Can Soc Sci 7(4):Article 4. https://doi.org/10.3968/j.css.1923669720110704.041
Zoomers A (2010) Globalisation and the foreignisation of space: seven processes driving the current global land grab. J Peasant Stud 37(2):429–447. https://doi.org/10.1080/03066151003595325
Zoomers A (2011) Introduction: rushing for land: equitable and sustainable development in Africa, Asia and Latin America. Development 54(1):12–20. https://doi.org/10.1057/dev.2010.97

Chapter 2
Military Frontiers and Neocolonisation: Land, Power and the Urbanisation of Lahore

Contents

Neoliberalism has also restructured social hierarchy by putting military at the top as influential capitalist class and the backbone of the national security apparatus. The 'authoritarian state' has formed with the help of five apparatuses i-e army, police, courts, bureaucracy, prisons to accumulate capital and power for military class and its foreign 'allies'. Harvey (2005)'s understanding of neoliberalism with regards to military is typified by 'cronyism', where corporations have close connections with military and provides them a route to military to access markets and take up infra- urban development projects. The alliance between the military and local and foreign capitalist classes play role in obtaining primary control over the urban land markets in country's major and big cities. DHA, Lahore has expanded into rural realm on the East-side in proximity to the cantonment for which it targeted agricultural lands and village communities to create high-end gated society by maxing out the real-estate value and to minimize resistance through reinforcing security infrastructure. It is only to cater the upper classes meanwhile poor and marginalised villagers are treated as burden. DHA ascertains its authority to design its own bylaws which were in violation with Lahore Master Plan, also has excluded itself from the requirement of LDA's approval for land development due to its legitimate position (the shared information has been gathered from literature, especially Ahmed (2023).

H. Javaid, *Land Commodification, Urbanisation and Social Change in Peri-Urban Villages, Lahore*, Advances in Asian Human-Environmental Research, https://doi.org/10.1007/978-3-032-10462-5_2

Abstract The urbanisation of Lahore is a story of both persistence and breakage, where agrarian landscapes give way to concrete under the long, dark shadows of empire and neoliberal globalisation. This chapter traces the historical trajectory of the city's expansion, situating it within colonial land regimes as well as the intensification of neoliberal reforms. While urbanisation is not new, in recent decades it has accelerated dramatically, framed as the engine of economic growth and modernization as a delusional marker of growth. Neoliberal policies, with their emphasis on deregulation and private capital, have created an insatiable demand for land. This transformation is not natural but actively produced through the collaboration of state and market - mediated by urban restructuring, land-use planning, and policies that enable eviction and displacement. On Lahore's periphery, agrarian fields are rapidly converted into real estate frontiers, shaped not only by speculative private capital but also by the state itself, acting simultaneously as regulator and development partner. By connecting colonial legacies with contemporary neoliberal urbanisation, the chapter argues that Lahore's growth is not merely spatial expansion but a political project of land, power, and control.

2.1 History of Urbanisation in Lahore

The history of Lahore is 2000 years old and gained much prominence during Mughals; however, dynamics of Lahore in terms of urban expansion gained shaped and momentum during colonial times (Abbas and Wakil 2023). Punjab was annexed by Britishers in 1849, and since then, it underwent massive transformation in the name of modernity/coloniality through land privatisation, developed agriculture, enclosure of commons and extraction of resources (Ranauta 2020) and most importantly Canal Colony Project in the region, with the establishment of extensive water irrigation network, which turned barren land into South Asia's one of the major centres for commercialised agriculture (Ali 1987). The agricultural colonisation through making canal colonies unbridled the process of urbanisation in Punjab (Singh and Singh 2019). The Punjab region was of immense importance to Britishers and a stronghold for British colonial power prior to independence and partition of sub-continent in 1947. The development of canal colonies with the establishment of canal-irrigation system during the late nineteenth century was not only to develop a commercial agriculture system in Punjab but also to erect the power structure which could support British rule in the sub-continent (Ali 1979). The conversion of uncultivated area or 'desert' into rich agricultural lands, the development of means of communication and transportation through railways, cantonment areas and cash crop production replacing local agricultural practices were not for the good of common man but to serve imperial rule and accumulation. It was exploitation of the natural resources of the region and ruination of the traditional economy (Singh and Singh 2019). The development of canal colonies led to massive changes in the physical and social landscape of the Punjab region. It created one of the biggest migrations from the more developed regions of East Punjab to the newly colonised lands in the West. For the colonists, it provided a huge workforce for large-scale commercial agriculture, and the migrant families inhabiting canal colonies sent their sons to serve the British Army (Ali 1988). Akhtar (2006) says that the development interventions gave British immense power and made it easier for them to get hands on the means of production such as land and water resources by allotting huge tracts of lands to select upper caste of landed agriculturalists and military personnel, thus creating a class of loyalists. Faction, which was of no use to the British Raj, was excluded from land rights and land grants such as the landless artisans and rural poor. It was the time when communities were divided into two major categories, i.e. land owners and the landless.

The bottom line is that the urbanisation of rich agricultural lands in Lahore is not a new phenomenon, as roots date back in nineteenth-century colonial rule (Glover 2008) as the new colonial developments of civil lines, cantonment, railway colonies and modern planned areas were developed outside the walled city. The direction of most post-partition development was towards the south, given the river Ravi on the west and the border on the eastern periphery. Most of the planned development was in line with the modern paradigm of grid-iron sprawl for primarily middle and upper class consumption (Chattha 2012).

2.2 Continued Colonisation

Even after gaining independence, the elites left by the British including military and civil bureaucracy have been taking the advantage of their position and accruing control over land and pegging down highest defence budgets by claiming the reward for their courage to protect the borders. Their involvement in politics of the country and alliances with landed class and feudal landlords mushroomed their strength, influence and power. Alavi (1972) explained Pakistan as post-colonial society (PCS amongst others) followed the inherited structure consisting of the organised bureaucracy and high-powered military structure which was established by British, i.e. the presence of powerful bureaucratic–military apparatus to rule indigenous social classes, and native bourgeoisie (once used to oppose military and landed classes) was not strong and apt enough to develop a new structure by themselves; hence, they couldn't dominate even in post-colonial period. On the other hand, weakness of one class gave more power and strength to military bureaucratic oligarchy to maintain and gather more power to be autonomous to rule the society.

The military not only became politically influential but also built an economic empire over the haul as they were involved in multiple economic activities; however, the military's interest in acquiring land to multiply its financial gains has become more evident through DHA's rapid expansion on the southeast periphery of Lahore in less than two decades. Not just for villagers but for DHA also, land is a symbol of power. Siddiqa (2007) states that the military is dominant when it comes to agricultural land as it controls 59% of the total land in rural areas. Further, she mentions that land which was allotted to military personnel from 1965 to 2003 in Lahore alone was 273,413 acreages (110,646 ha), and the built-up area by DHA in Lahore is only 30,000 acres (12,141 ha).

2.2.1 *Military's Real-Estate Venture: Defence Housing Authority (DHA)*

The armed-forces real-estate business through developing gated housing schemes known as Defence Housing Authorities (DHA) has been one of the most profiteering ventures. DHAs have been built in major cities of Pakistan such as Lahore, Karachi, Islamabad, Multan, Bahawalpur, Gujranwala and Peshawar (Article 2023). It has been considered the most up-scaling gated community which is not just a residential area for affluent but known as a brand in housing. DHA is a military real-estate business and one of their ventures through which they have built their economic empire, and uncontested political position gives them power to 'rule'. DHA's ideology in designing residential communities is the continuation of colonial practices that are and will only exacerbate social and economic differences (Shahbaz 2022). DHA has been expanding at the expense of rural areas from where the authorities can easily acquire land from villagers at the lowest possible rate to create huge profit margins. The acquisition raises questions on 'choice', fair deals, compensations, use of violence and coercions and several unjust tactics which label land tycoons as feudal landlords. (Land acquisition process in contemporary times will be discussed in detail in result section.)

For instance, Akhtar and Rashid (2021) take Islamabad as a case study to foreground the processes of urbanisation, fuelling dispossession of urban and rural working classes living on agrarian–urban frontier to lubricate windfall profit-making business activities through real-estate development due to land speculation under 'financialisation', which is a common practice in several other places in the world. Authors argue that the city and its development in the country are highly militarised, draw a linkage between militarisation and urbanisation and primarily usher to unregulated suburbanisation and inappropriate utilisation of land. Military has taken over 12% of state land, of which 62% alone from Punjab. Since Pakistan's military is the country's largest gated housing scheme developer (Rahman 2016), it has been acquiring land through Land Acquisition Act, 1894, which helps military to have financial gains that often result in displacing communities. As the military has been involved in the politics and has maintained privileged political clout, it gives an easy access to details concerning city development and planning which in turn becomes more profitable for real-estate developers (Siddiqa 2007). The removal of Multan mango orchards for the DHA (Batool 2021; Staff 2022) was another manifestation of the cruel and dismal side of the development by the most influential realtor who is also the face of the State.

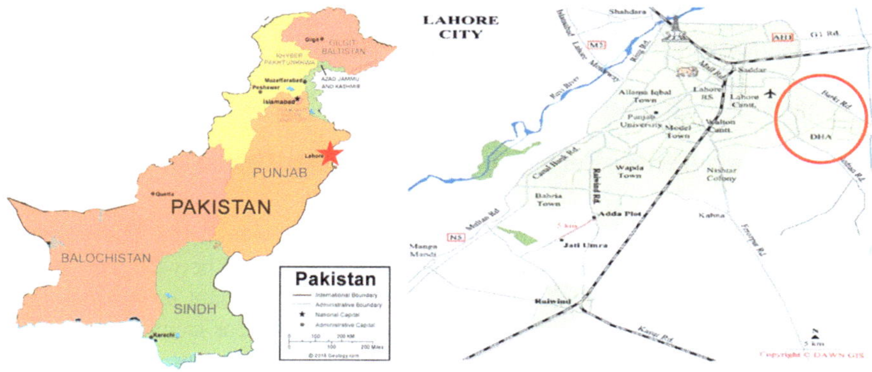

Fig. 2.1 Introduction to Research Site

2.2.2 *The Power of DHA in Lahore*

The army announced the launch of its scheme for developing housing society for retired army officers in the vicinity of Charrar village in 1977 (see Fig. 2.1). While initially, Charrar residents were reluctant, yet they sold the land as the prices were sufficiently high for the small, cash-poor farmers. Dwellers used the money to buy larger parcels of land farther to the east, but kept their houses. All the land except the settled land and graveyards were acquired by the then Lahore Cantonment Cooperative Housing Society (LCCHS) which has since 2002 been restructured as Defence Housing Authority (DHA). The first four phases of this housing were developed in 15-year time span from 1978 to 1993 (Zaman and Baloch 2011), and then, it took unimaginable pace from early 2000s till date. The development by DHA is underway at bewildering speed erasing historical villages, fields and orchards. Due to the proximity to the Indian border, the need for approvals by the military for growth on the eastern periphery of Lahore ended as the army itself became a major land development agent in the area. Today, DHA Lahore comprises thirteen phases and spans approximately 44,000 acres (17,806 ha) of land across the city (Plopterty, n.d.). The acquisition of land for the later phases intensified around 2002 along *Barki* and *Bedian* roads, with built-up development beginning in 2006–2007. Earlier, DHA Phase 5- launched in 2001 - had already marked a major expansion by acquiring large swathes of agricultural land in the southeastern part of Lahore. (Anjum and Hameed 2007). The expansion continues and does not seem to come to an end.[1] To acquire land, the 'Land Acquisition Law 1894' has not changed

[1] The personal interview (2024) with well-established land providers who work for DHA revealed that development-related future plans of DHA will cause BRB canal and *Hadiara* drain to be removed. Also, the military farmlands across *Barki* will be erased as DHA is aiming for relentless expansion.

much since the colonial era, according to which developer has to pay compensation to the landowner.[2]

2.3 Research Site: Peri-Urban Villages Southeast of Lahore

Southeast of Lahore is the latecomer in the process of development. The periphery took a long time to get urbanised as the cantonment is situated between the city and the border area villages. Border itself constrained development as it was a huge reservation for people due to geo-political concerns and fear stemming from the rocky relations between two countries, due to which the creation of a market in land for housing was on halt. Later, all masterplans for Lahore concentrated on the south-eastern periphery for development which was characterised by suburban sprawl. It was only in the masterplan (LDA, 2016) for Lahore that agricultural area around *BRB Canal, Bedian Road, Hudiara Drain, Khaira distributary, Barki Road* and *Shalimar Link Road* was designated for development (Javed 2015).

The southeastern periphery of Lahore has been undergoing massive transformation since 2000. The change in physical landscape took on a hectic pace with the construction of the new terminal of the *Allama Iqbal* International Airport opened in 2003, adjacent to the old terminal on the southeastern fringe of the Lahore city. This former rural area is dotted by dozens of villages, which till 1947 lay sandwiched between the twin cities of Lahore (Pakistan) and Amritsar (India) which are 2500 km^2 apart. The area on the eastern outskirts of Lahore is not new to drastic social change. Lying close to the border between Lahore and Amritsar, the villages were thrown into the turmoil of population exchange in 1947. Historically too as the villages between two major cities of Punjab, they had been very prone to impacts of urban political and economic change compared to the villages deeper in the hinterland. However, recent social change is more structural. Even during partition in 1947, villagers had moved as clans; they were mostly from villages sharing links through marriage and exchange, and their basic way of life was tied to the land. The twin cities of Punjab, Lahore and Amritsar, bore the brunt of 1947 partition the most (Talbot 2007).

Today, the international border between India and Pakistan divides this area almost in half. Two major roads *Barki* and *Bedian* connected towns before the partition of British India. This is a rich agricultural land watered by 117 km^2 long *Bambanwala–Ravi–Bedian* (BRB) canal, also called *Ichogil* Canal which runs on the eastern boundary of Lahore and its distributaries, built in 1948 by locals for

[2] DHA is private housing scheme and LDA is the Government or State's controlled urban authority responsible for land development in Lahore. LDA was supported by the Punjab Acquisition of Land (Housing) Act 1973, which was enacted during the Bhutto's reign. It helped state to take over land at affordable rate which was 40,000 PKR (133 €)/0.405 ha until Zia-ul-Haq came and not only cancelled the Act in 1985 but replaced it with Land Acquisition Act of 1894, which was a huge set-back for LDA (Ahmed 2023; Anjum and Hameed 2007).

defence purpose to ensure safety of the city (Khan et al. 2015), which too flow seamlessly across the border. These villages feed the city as agricultural lands have a capacity to produce variety of food such as vegetables (e.g. carrot, peas), rice, wheat, maize, mustard, sugarcane and fruit trees (such as mango, orange and guava, amongst others), and they are also a source of manpower for the city.

The class structure in the rural realm is attached to land itself. As Ahmad (1970) and Gazdar and Mallah (2012) explain, rural stratification is not only on the basis of agrarian classes but also organised in a social order according to the occupations (agricultural and non-agricultural), socio-economic status (possession of cattle, income generated from the harvest, ornaments, utensils, furniture, tools), asset ownership and in Ahmad (1970)'s word, 'two broad division on the basis of occupation are cultivators (*Zamindar qoum*) and non-cultivators'. Nevertheless, there are further sub-divisions within these major groups as well. Big landlords are at the top of the social hierarchy who themselves never cultivate land but hire others to work for them. Then come rich peasants, who work themselves but hire others also. Middle peasants are only dependent on their own labour and do not recruit others. Poor peasants are those who are landless but still belong to the category of cultivators as they work on other people's land. Finally, artisans also work as agricultural labour and service providers to the village. Rich landlords are the ones who control means of production which gives them power and influence over the others (Hussain 1984). Authors might have looked at the class structure differently, but one pattern remains unchanged, i.e. the people at the lower rung were treated as disposable, and their weak positions only weakened further since British times, and the cycle continues. The green revolution turned big landlords into capitalist landlords, and middle peasants or small landholders had to give up on their land and had to be totally at the mercy of landlords which increased the exploitation. Land reforms which were introduced to resolve the issue of unequal distribution of land never worked out and failed to protect the landless and working class due to political and power clout of the big players (Mughal 2019).

During Mughal era, there was a protection to peasants and even the landless which was coming from a rural structure in place. The landlord and tenant/share cropper had co-dependence on each other as all were tied to land as means of production. Grover (1963), in his paper discussing land rights of various classes in Mughal era, quotes Jahangir who issued order that *ryots* (peasant or tenant farmer) could not be expelled against their will, neither did they settle men and cattle from other villages in his own Jagir; loans must also be provided to increase agricultural yield and to bring wasteland under cultivation. Further, he states that in the Mughal Age, the state never claimed absolute ownership or farmlands and recognised private property in land. Moreover, the hereditary rights pertaining to land ownership, its sale, purchase, mortgage and transfer were bestowed upon *Zamindars* (agricultural colonisers), as even the tenants working on the land had also the hereditary rights. The people on the lowest rung of social ladder such as landless who were hired for cleaning jungle or as seasonal labour during the harvest were given the share in the produce according to existing customs and rules in a village. Most strikingly, during the Mughal epoch, there was no serfdom or 'villeinage' that would atrociously tie peasants to land and feudal lords. In addition to that, land was not even taken as a commodity the way it started happening in the late eighteenth and nineteenth centuries. British invasion in the sub-continent not only broke the very structure that used to provide shield to every class despite its shortcomings and corruption but also introduced redeveloped structure which was meant to support a higher class of feudal lords to acquire patronage-brought landless, artisans and small peasants to the sheer exclusion and marginalisation by piercing safety blanket. Even as the British Raj left the sub-continent, colonialism never loosened its grip even in the post-colonial era.

2.4 Historical Significance Attached to Villages in the Research Area

In keeping with the broad land holding profile of Lahore district (GOP 2011), small landholdings and intense cultivation define the local ecology of the eastern periphery of Lahore city. It is a community of villages connected both socially and physically. Shrines dot the landscape and are sites of annual fairs frequented by villagers from across the area, and communities are linked by caste and marriage relations. The interlinked water courses, natural drains and shared road connections to Lahore bind the wide plain as does a web of social relations.

Most villages are historical and have been inhabited for a few hundred years[3]; however, some are smaller settlements of dozens plus homes, where people usually who moved in clans are residing. Many villages, such as *Padhana, Charrar, Heera Singh* and *Lehna Singh,* were dominated by big non-Muslim landlords prior to 1947. The land of the non-Muslim was distributed to migrants from India; the *Meo* community from *Rohtak* and *Karnal* were one of the notable communities who came and took over the land vacated by non-Muslims. Many of the migrants were Muslims who lived close to the border on the Indian side and had historical and familial ties with these border villages.

Majority of the villagers were small farmers, but there were a large number of landless, artisanal castes, poor Christians and Muslim Sheikhs (lowest caste Hindus who had converted to Islam). Muslim majority villages were poorer than those with Sikh and Hindu landowners, for example, the village of *Mandianwala* has virtually no migrants, but amongst the Muslims, there was no big landlord. Villagers in some villages such as *Mandianwala* and *Barki* considered their relative equality in poverty, the basis for their strong community spirit. In *Barki* village, a local lawyer remembered fondly the enthusiasm with which the villagers had voted the populist politician *Zulfiqar Ali Bhutto* to power in the 1970 election. He felt that there was no significant abuse by police and violent crime typical of other parts of Punjab where

[3] Personal correspondence with local historian Iqbal Qaiser who is compiling the histories of villages of Lahore district based on the descriptive narratives of the village census in 1881.

there are powerful feudal lords. Large parcels of land in few villages are the military farms, and a large acreage belongs to an important Sikh temple, *Gurdwara Dera Chahal*, which is administered by a special government department, the Evacuee Trust Board. People cite the influence of the military and the absence of feudal culture as a reason for a relatively crime-free environment and security in the area. *Dera Chahal* has gone through a visible change from an area surrounded by agricultural lands to asphalt roads, renovated houses and urbanised living where the façade does reveal mixed-class residential area.

Pakistan was once known as an agricultural-based country; now, it is known as a 'transforming' country according to a World Bank report based on the relative contribution of agriculture to economic boom and employment (Dudwick 2011). Dudwick also combines 'the poor' and geography of the city that is economically behind 'rural areas' to examine the rural–urban transformation in South Asia. The purpose of giving this reference is to re-draw attention towards the southeastern periphery of Lahore, especially since the time when in 1999 Lahore Cantonment Cooperative Housing Society (LCCHS) was set up, as the name signifies 'military lands', which is now DHA phase 1. DHA took another jump when it was finally federalised in 2002. Then, the creation of the terminal of *Allama Iqbal* International Airport in 2003 under the rule of *General Pervaiz Musharraf* opened floodgates for land acquisition for the infrastructure development. The infrastructure at such a large scale opened the floodgate for setting up a robust market in land, and the presence of the military in the area provided a certain amount of security to people who felt encouraged to invest. This peripheral area which was connected to the main Lahore city through major road networks, later by the ring road, was the agricultural area and the suburbs of 'the poor'. People used to commute to the city for work and travel back to their homes in villages. The soil was so fertile that it could grow any seed a farmer would sow. At least one of the family members did not give up on farming; people with small landholdings might not get the higher returns but used to produce for the household consumption that provided them freedom from market. The houses on the periphery were usually big with a space to keep the animals. The common businesses in these villages were farming and milk selling which used to go hand in hand. Labelling a place as 'poor' only on the basis of Gross Domestic Product (GDP) does not do justice. They were cash-poor but rich in resources and socio-ecological assets. However, this wealth does not fall in mainstream economy but has become an agenda when it comes to plundering and robbing the place of its resources and villagers' social capital. Villages one after the other were subsumed in developing a city for the wealthy faction of society at the expanse of those who were treated as mere cog in the wheel. DHA being the biggest actor started acquiring agricultural lands through brokers by keeping the business under the wraps. The presence of military and its business in land became rife through physical manifestation on land-use change. The most recent development is the building of cavalry ground underpass, *Ghora chowk* flyover, *Bedian* underpass and Walton flyover to provide signal-free commute to the DHA residents that make 1.5% of total Lahore's population for which Punjab Government spent PKR 7 billion or 23,144,168 € (twitter, 2024). In the past two decades, the pace of urbanisation transcended the

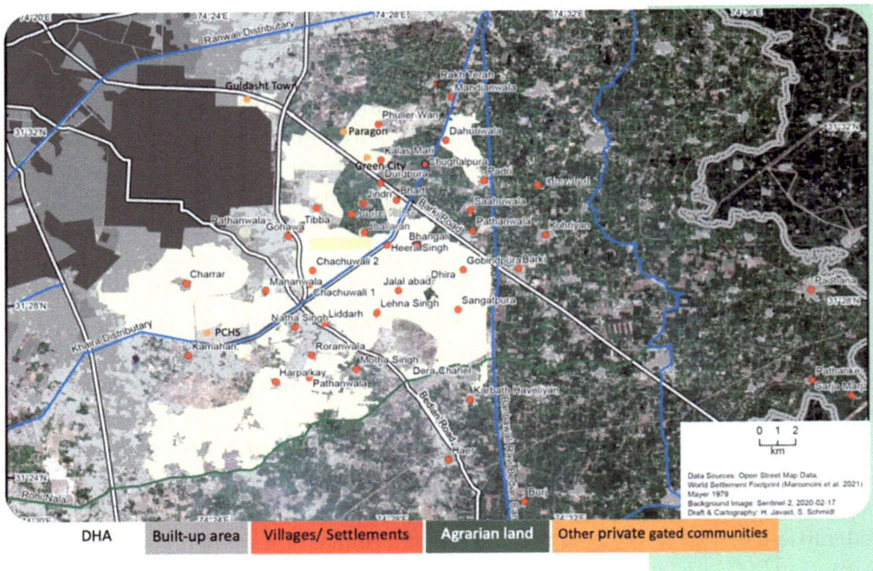

Fig. 2.2 Footprint of Defence Housing Authority (DHA), Lahore

period of change happened in Lahore in the past two to three centuries alone. The virtual representation of physical transformation (see Fig. 2.2) is appalling as it clearly shows the infrastructure spreading over the terrain like a fire. It shows a dichotomy even within the limited region between urbanised city and green agricultural areas towards the east.

Where Fig. 2.1 shows expansion of Lahore over the period of time, starting from Mughal era as Lahore held a special place for Mughal emperors. With the arrival of British Raj, and the establishment of canal colonies, the city transformed tremendously. The development continued after partition which was (as mentioned earlier) south-ways over the long haul. However, neoliberal epoch unleashed irregular unchecked city expansion also towards east, where DHA has been the primary player (see Fig. 2.2) but has opened floodgates for many private developers and investors. The pace of urbanisation is unimaginable. Lahore took couple of centuries to be transformed, but the southeastern periphery took only 20 years for rapid urban expansion and commodification of arable lands and causing havoc in localised communities' agrarian world. The DHA's 40-year journey (1977–present) in totality when military started acquiring land from Charrar and getting bigger and powerful with passing years (see Fig. 2.3) pertinent to building high-end gated communities emerged as the most influential developer that claims to be answerable to

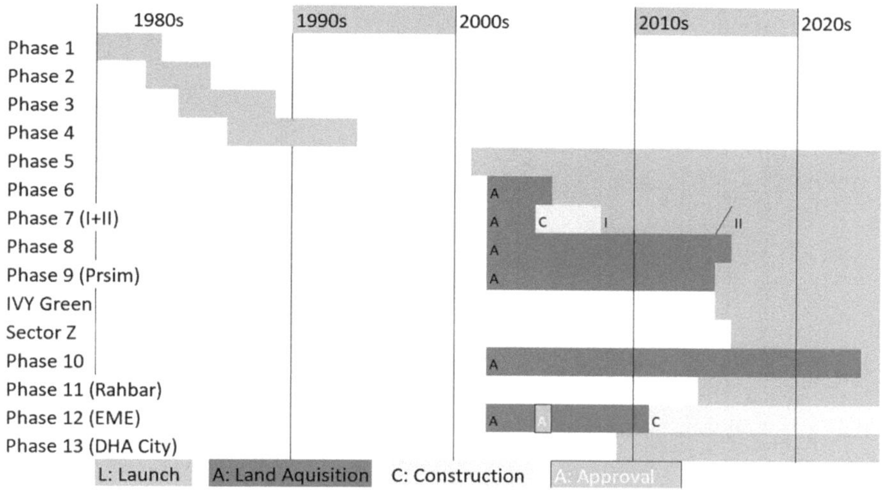

Fig. 2.3 Years of military's real-estate business in Lahore

none. Otherwise, over 25 years (1999[4] to date) of military's established economic empire of which DHA is a big part, the Phase 10 has just been launched and announced in February 2024.

A list of villages follows, including those that have been removed and those the researcher was unable to visit (Jalal Abad, Sangatpura, Gobindpura), to maintain inventory and to document how development-driven urban expansion continues to eliminate settlements. Nonetheless, the purpose of adding these villages to the list is an attempt to develop an inventory of villages which were and are present in the area (see Fig. 2.4). The villages that still remain under the pressure of elite housing construction are undergoing transformation: some are on the verge of extinction, others have become pocketed or walled off by DHA phases or other private housing schemes, and some have turned into examples of degraded ecology and ghettoization. The villages are divided according to the development stage these locales are at:

(a) **Razed Villages:** These are the villages that are removed from the face of the earth completely and supplanted by the symbols of modern infrastructure.

- **Completely razed** (7): Pathanwala, Heera Singhwala/Heera Singh, Gobindpura, Jalal abad, Sangatpura, Haveliyan, Luhar Pura.

[4] 1999—when LCCHS finally converted to DHA (https://dhalahore.org/about-us/history/)

© Google earth

Fig. 2.4 Old and New Landscapes

- **On the verge of removal** (in total 3): Chak[5] Dhira, Jhallaran, Natha Singh.

Note: The DHA phase 8 masterplan indicates the removal of Chachuwali in near future.

(b) **Settlements**: Agricultural lands are sold and houses are left behind; hence, they are now no longer considered as villages. Also, the removal of arable lands and the urbanism-takeover in the localities are conspicuous (in total 26): Jindra Kalan (a part of it has been taken over by DHA), Chak Bhart, Lehna Singh, Dahoorwala/Dahuriwala, Kohriyan, Harpalke, Kalasmari, Bhangali, Chachuwali 1 and 2, Tibba, Mananwala, Roranwala, Liddarh, Gohawa, Mandianwala, Haer, Charrar, Saduwala, Ghawindi, Kamahan, Mota Singh, Phuller Wan, Dera Chahal, Durgpura, Barki.

(c) **Villages** (in total 9, including border villages and military farm villages): Burj, Chughalpura (Habib abad), (Rakh) Padri, Padhana, Pathanke, Sarja Marja, Jahman, Jindri, Haveliyan/ Karbath Haveliyan.

Note: Border villages are distinctive as they are farthest to the east and in close proximity to the border; they cannot be invaded by development (for now) due to defence purposes; similarly, tall trees are not allowed to be planted for the same reasons. These villages are put under same category as they are still having agrarian living, and the rural culture is still intact. Yet, the border villages are taken as a reference point.

[5] Chak refers to the villages which were established in canal colonies under the British Raj.

(d) **The gated communities to which people migrated from villages[6]:** Punjab Co-operative Housing Society/Al-Ameen Housing Society,[7] Guldasht town, Paragon city, New Abadi Barki.[8]

Note: The above-mentioned gated communities are not the only ones where people from villages migrated or new settlers came from other places to live in those communities. There are many and soaring with time. However, the people I have interviewed were from these newly formed societies.

[6] After selling agricultural lands, people bought agricultural lands to the farthest; who sold houses either migrated to other villages where they could have a similar environment or the ones who had better resources moved to newly built housing societies. As moving to DHA is not a cup of tea for villagers, not everyone can move to high-end locale that is already labelled as high-class, posh or elite. It is not about the financial affordability but affordable living in terms of psychological and social context too. Buying a house and a luxury car does not allow these migrated villages to fit in the society that is unable to look at them as equals. In order to make their place, people are supposed to carry themselves in a certain way that includes spoken, branded outfits, latest material gadgets, etc. It demands people to lose themselves in the process of becoming an acceptable version for the upper class which is a mere fallacy. Therefore, people tend to move out to such gated communities where they would not feel out of place and can find people who are coming from the similar backgrounds.

[7] However, these gated communities also have a reflection of class segregation; it might not be as stark as compared to DHA, but it does exist. This gated community is close to DHA phases 3 and 5 and almost at the back of the latter. It is comparatively a small society with 200–300 houses with clean roads. In terms of *braderi*, there is mixed population such as Meo, Jatts (who actually own this place before society was built) and army officials also have plots in the society. Jatts usually have a large landholding, and the Meo families (personal interview) were amongst the three families of Lehna Singh who had good amount of land, sold it to DHA, bought farmlands in Haer and joined property business. It can be characterised as a locality for the middle class. Whereas people from Kalasmari preferred Paragon, and one of the reasons lie in the proximity, so that after moving out, they could stay close to their village community. Paragon, Green city, Punjab Society can accommodate the middle class.

[8] However, it is different from the above-mentioned. People who moved here were coming from Sangatpura or Nathoke, where they did not have the agricultural land, and residential area was also quite small. Men in the families were working as wage labourers. The landscape of the society was not close to above-mentioned gated communities, and people from lower-income groups were residing there.

References

Abbas M, Wakil I (2023) Exploration and analysis of the drivers of urban sprawl in Pakistan: a case study of Lahore—RASTA. RASTA 66. https://rasta.pide.org.pk/cgp/exploration-and-analysis-of-the-drivers-of-urban-sprawl-in-pakistan-a-case-study-of-lahore/

Ahmad S (1970) Social Stratification in a Punjabi Village. Contrib Indian Sociol 4(1):105–125. https://doi.org/10.1177/006996677000400107

Ahmed A (2023) The rise of military capital in Pakistan: military neoliberalism, authoritarianism and urbanization. Geoforum 146:103846. https://doi.org/10.1016/j.geoforum.2023.103846

Akhtar AS (2006) The state as landlord in Pakistani Punjab: peasant struggles on the Okara military farms. J Peasant Stud 33(3):479–501. https://doi.org/10.1080/03066150601063058

Akhtar AS, Rashid A (2021) Dispossession and the militarised developer state: Financialisation and class power on the agrarian–urban frontier of Islamabad, Pakistan. Third World Q 42(8):1866–1884. https://doi.org/10.1080/01436597.2021.1939004

Alavi H (1972) The state in post-colonial societies: Pakistan and Bangladesh. New Left Rev I(74):59–81

Ali I (1979) The Punjab Canal Colonies, 1885–1940 [The Australian National University]. https://openresearch-repository.anu.edu.au/handle/1885/112646

Ali I (1987) Malign growth? Agricultural colonization and the roots of backwardness in the Punjab. Past Present 114(1):110–132. https://doi.org/10.1093/past/114.1.110

Ali I (1988) The Punjab under imperialism, 1885–1947. Princeton University Press. https://www.jstor.org/stable/j.ctt7zvvct

Anjum GA, Hameed R (2007) The dynamics of colonization of peripheral housing schemes and policy options in case of Lahore. Pak J Eng Appl Sci. https://journal.uet.edu.pk/ojs_old/index.php/pjeas/article/view/214

Article E (2023) Defence housing authority. Wikipedia. https://en.wikipedia.org/w/index.php?title=Defence_Housing_Authority&oldid=1131887613#cite_note-1

Batool Z (2021) Horrible! Why DHA has cut about 6,000 acres of mango orchards in Multan? Parhlo. https://www.parhlo.com/6000-acres-of-mango-orchards-cut-by-dha-in-multan/

Chattha I (2012) Economic change and community relations in Lahore before partition. J Punjab Stud 19:193–213

Dudwick N (ed) (2011) From farm to firm: rural-urban transition in developing countries. World Bank

Gazdar H, Mallah HB (2012) Class, caste and housing in rural Pakistani Punjab: the untold story of the five Marla scheme. Contrib Indian Sociol 46(3):311–336. https://doi.org/10.1177/006996671204600303

Glover WJ (2008) Making Lahore modern: constructing and imagining a colonial city (NED-new edition). University of Minnesota Press. https://www.jstor.org/stable/10.5749/j.ctttskzh

GOP (2011) "Punjab development statistics, 2011" Bureau of Statistics, Punjab, Lahore. Government of Punab

Grover BR (1963) The Indian economic and social history review: nature of land-rights in Mughal India. Indian Econ Soc Hist Rev 1(1):1–23. https://doi.org/10.1177/001946466400100101

Harvey D (2005) A Brief History of Neoliberalism. Oxford University Press https://doi.org/10.1093/oso/9780199283262.001.0001

Hussain A (1984) Land reforms in Pakistan: a reconsideration. Bull Concerned Asian Sch 16(1):46–52. https://doi.org/10.1080/14672715.1984.10409782

Javed U (2015) Rezoning Lahore. DAWN.COM. http://www.dawn.com/news/1160868

Khan M, Javeed F, Anbrine S (2015) The impacts of urbanisation on avian communities of LAHORE. J Res Archit Plann 18:23–32. https://doi.org/10.53700/jrap1812015_3

Mughal MAZ (2019) Rural urbanization, land, and agriculture in Pakistan. Asian Geogr 36(1):81–91. https://doi.org/10.1080/10225706.2018.1476255

Plopterty (n.d.) DHA Lahore [https://plotperty.com/pakistan/lahore/dha-lahore]. https://plotperty.com/pakistan/lahore/dha-lahore

Rahman T (2016) Enabling development: a housing scheme in rural Pakistan. University of Oregon. http://hdl.handle.net/1794/20410
Ranauta J (2020) Transnational modernity/coloniality: linking Punjab's canal colonies, migration, and settler colonialism for critical solidarities in Canada. Stud Soc Justice 14(2):Article 2. https://doi.org/10.26522/ssj.v14i2.2272
Shahbaz S (2022) Deconstructing DHA Lahore: analysing post-1980s Military operated housing through three spatial-ideological systems [Master Thesis, University of Waterloo]. https://uwspace.uwaterloo.ca/handle/10012/18123
Siddiqa A (2007) Military Inc: inside Pakistan's military economy. https://readersend.com/product/military-inc-inside-pakistans-military-economy/
Singh B, Singh B (2019) Punjab under the British rule: historicising the local transformations. Indian Hist Rev 46(2):207–226. https://doi.org/10.1177/0376983619889520
Staff S (2022) Approx. 100,000 Mango trees cut down to pave way for DHA Multan. Siasat.Pk News Blog. https://blog.siasat.pk/approx-100000-mango-trees-cut-down-to-pave-way-for-dha-multan/
Talbot I (2007) A tale of two cities: the aftermath of partition for Lahore and Amritsar 1947–1957. Mod Asian Stud 41(1):151–185. https://doi.org/10.1017/S0026749X05002337
Zaman K, Baloch AA (2011) Urbanization of arable land in Lahore City in Pakistan: a case-study. Can Soc Sci 7(4):Article 4. https://doi.org/10.3968/j.css.1923669720110704.041

Chapter 3
From Satellites to Sacrificial Zones: Rewriting 'Development of Underdevelopment' in Lahore's Urban Frontier

Contents

Abstract This chapter reconsiders Andre Gunder Frank's thesis of the 'development of underdevelopment' through the lens of Lahore's urban frontier. Once embedded in agrarian systems as productive 'satellites', peri-urban villages have become sacrificial zones for speculative real estate and state-backed urban expansion. Drawing on ethnographic research from 2014 to 2022, the chapter traces how fertile farmland and communal life are dismantled through land commodification, dispossession and the advance of Defence Housing Authority (DHA) projects for expansion of gated community for the elite. The analysis reveals both visible and invisible transformations unfolding in agrarian peripheries. Material changes include farmland loss, declining water and food quality, and broader ecological degradation. Equally significant are the less perceptible disruptions, such as the erosion of trust and kinship, emotional and psychological dislocation, and the ongoing

H. Javaid, *Land Commodification, Urbanisation and Social Change in Peri-Urban Villages, Lahore*, Advances in Asian Human-Environmental Research, https://doi.org/10.1007/978-3-032-10462-5_3

struggle to preserve agrarian memory. Together, these shifts illustrate how environmental, social, and affective worlds are being reshaped, leaving communities suspended between a disappearing past and an uncertain future. This is where I am extending Frank's insights, as it shows how dependency is reproduced not only globally but also locally in the lived landscapes of peri-urban Lahore. Development here emerges not as progress but as a political project of fracturing, where life and land are reordered amid dispossession and contested futures. It further reinforces that development is not neutral, but a designed, project - a part of global agenda is intrinsically political, and it's not meant to improve a lay man's life but a *discourse* tied to power to serve big shots.

3.1 Introduction

The research focuses on socio-spatial transformation due to rapid urban sprawl and attendant environmental disruptions on the periphery of Lahore. It is a documentation of conversion of long sustained village settlements into degraded socio-ecological realms. It is a holistic study aimed at analysis and theorisation of processes of social and material change in the wake of land commodification—the creation of a market in land. Land in the villages was commodified once powerful cash-rich investors appeared on the scene to acquire land for development. The cash compensation for land was accompanied with the force of policy and institutional support of the state in favour of land transfer for elite housing. Dispossession, displacement and loss of agrarian land were justified by the award of cash compensation to the landowning farmers. The theory of urban political economy (UPE) has been a holistic, critical and very productive theoretical framework to unfold the actors, institutions and power dynamics, that are not only in terms of social and political position of certain actors but the power and influence that come from money to develop areas for profit (Swyngedouw and Heynen 2003). According to Swyngedouw and Ward (2020), neoliberal development on the periphery, which begins with large-scale land grabbing, is nothing less than a venture for creating 'mass proletarisation'. UPE was christened as such by Eric Swyngedouw in 1996 but has deeper roots in Marxist political economy (Heynen 2017). David Harvey and others (Harvey 2001; Kugkhapan 2014) theorise the process of land development for urban expansion as a spatial fix strategy for capital. Therefore, the 'question' of geography and its politics is of immense importance in this trajectory of capitalist development, not only how it alters space or reproduces but also how unique geographies provide ground for capitalism to grow through uneven development (Smith 2008). In present times, such development is not only destroying environmental resources but depleting social capital and dismantling traditional societies of global South. Narrative or storytelling in urban world contributes in forming perceptions, which has an impact in forging reality (Ameel et al. 2023). Notions attached to even or uneven and developed or undeveloped vary with 'who' speaks for 'whom' and 'how' one narrates the story of change. Rural communities have been living through a process and become weaker and poorer (in the literal and figurative sense) as a result of state policies aimed at investment in new developments creating landscapes for upscale lifestyles of the rich and erasure of rural locales with local production and subsistence.

Importantly, the most distinct character that drives phenomenon like urbanisation is ‘accumulation’ not only through ‘dispossession’ in economic terms and tangible assets (such as land, home, means of livelihood), but this study is looking at dispossession on much deeper but underrated yet fundamental aspect of change, i.e. the dispossession of intangibles that make a quality of life. The aim of this study is to make visible the invisible. The villages that are lost in the process used to carry history and its own identity like all those villages who are still standing in precarious conditions, putting people and communities in constant state of anxiousness with a fear of losing sense of belonging and identity to a place, setting the displacement clock which might never stop ticking (Celestina 2016). The communities are at warfare where they could see cash ruling. Also, the development changes the fate of rural–cultural subsistence finally to be drawn into the market economy and then be lost in the labyrinth of monetary gain and loss.

3.2 Relevant Theories and Concepts: Building Theoretical Framework

Political ecology as a field has attracted scholars from various fields such as anthropology, development studies, environmental science, sociology, geography and so forth. Despite varied backgrounds, researchers tend to learn the relation between social, economic and political dimensions to understand the environmental and ecological change. This term was first introduced by French philosopher Bertrand de Jouvenel in 1957 and anthropologist Eric R. Wolf in 1972.[1] In the 1970s and 1980s, it evolved as an approach for studying change especially in the so-called developing countries. This research has started looking at the process of urbanisation through the lens of political ecology and has been put in the Marxist’s urban political ecology framework.

According to Swyngedouw and Heynen (2003), it is a holistic view of the connected socio-political (power structures) and economic processes with a historical understanding of ecological/material world space as a dynamic living environment. Rapid urbanisation is a limb of ‘development’ across the Global South creating large-scale commodification of peri-urban and rural land under a neoliberal capitalist development paradigm. The neoliberal economic model has imposed free markets and promoted privatisation, resulting in massive enclosure of public and private land (land grab/ global land rush). It has brought in its wake displacement, dispossessions, loss of livelihoods, plundering of natural resources, food insecurity and

[1] http://www.ejolt.org/2013/02/political-ecology

environmental and ecological degradation. The 'development' narrative of world class cities presents this as introduction of modern infrastructure of road networks, airports, industries, high-rise buildings and secure (gated) communities. Gross sub-urbanisation on the urban peripheries of cities such as Lahore, Pakistan, is seen as path to unburdening the historical inner city, provision of employment opportunities, influx of cash for poor farmers and a gateway to better life (based on increased consumption). Therefore, UPE can allow to acknowledge the process of 'urbanisation' as a concoction of various underlying processes such as political/role of a state, coercion, violence of power (termed as extra economic forces), economic models and policies, and social and ecological which all contribute in making urbanisation as a leading process for creating unequal landscapes and inequitable societies with the motive of capital production in cities (Cornea 2019). The present-day development and the disposal of communities are seen through David Harvey's theory of accumulation by dispossession where most of wealth or resources go under the hegemony of a few by depriving the masses. This further leads to his another theory called spatial fix, according to which capitalism needs to find places to dump surplus capital to avoid over-accumulation (Kugkhapan 2014), which resultantly creates uneven development (Khan 2020; Smith 1982, 2008).

However, urban political ecology is focused on 'city' unit not only for analysis (Brenner 2014), but the 'methodological cityism' has excluded other aspects of urbanisation processes (Angelo and Wachsmuth 2015) happening spatially. Urbanisation is not just an urban form or phenomenon but a process of capital accumulation that is not limited to cities only but has expanded to rural areas which carry a different ecology all together. Glover (2021) shared a part of his personal communication with S. Gururani regarding 'agrarian urbanism' (Gururani 2020) that much of the urbanism in India is more deeply rooted in agriculture and the agrarian mode of production burgeons relations of land and community; hence, urbanisation is no longer a city phenomenon but is ensnared in 'rural processes and relation'. This is comprehensible that why agrarian question has caught attention again, and why it is pivotal to combine agrarian and urban studies together to have deeper understanding. As the transformation on the agrarian–urban frontier can't be grasped unless it includes land question which the 'urban studies' especially centring on GN do not apprehend with same urgency. This is one of the reasons that why south Asian urbanisation driven by changed agrarian dynamics cannot be seen from the same lens as of the GN (Gururani and Dasgupta 2018).[2]

[2] Urbanisation is certainly not a new process, but the urban transitions occurred in GN after 1800 with industrialisation and second industrial revolution during nineteenth century, whereas in GS, the urbanisation took pace since 1950 or even later (https://ourworldindata.org/urbanization). However, the period of the 1980s onwards—the era of neoliberalisation—has increased the process through structural reforms, land markets, FDIs, infrastructure building and unimaginable urbanisation for more consumption. The GS urbanisation is accelerating and distinctive due to change in agrarian mode of production (Gururani and Dasgupta 2018). In Pakistan, there is a need to have more explicit and empirical research on rural land and communities and how state policies and power relations create havoc. Nausheen Anwar highlights important processes leading agrarian–urban transformation as a deeply political process.

Social relations are embedded in spatial relations and deeply intertwined, whereas capitalism has reduced everything to exchange value in a way that money is ruling all the aspects of life (Moore 2015). In the villages, studied for this research, the injection of money is de-stabilising and impoverishing the communities. The villagers of peri-urban Lahore whose agricultural lands were acquired for a price had lived fairly independent of monetary exchange for many essentials of their lives, both material and intangible. They were cash poor but rich in many aspects. Conceptually, it can be compared to Frank (1966)'s 'the development of underdevelopment' which argues that countries are not underdeveloped due to lack of potential of growth, but the First World became powerful and wealthy by keeping other nations underdeveloped for their vested economic and political interests. The borrowed development theories and policies from the GN state that development happens in stages and underdeveloped countries are labelled as situated in early stages of development. This is done by rejecting their past social and economic history based on exploitation during colonisation. According to Frank (1972), this unequal relation between core and periphery has also created a class of elites and dysfunctional social, political and economic structures enmeshed in capitalist relations; further, he used India as a classic example to explain the process of fostering unequal development through trade and socio-political scenario. The same analogy is used on the parasitical nature of relation of urban areas on the hinterland or traditional societies which are not backward, but the modern enclaves and the growth of a ruling class make them 'developed' at the expense of the hinterland. Hence, capitalist neocolonisation has rekindled the history not on the global level but within the city and its dependence on the countryside for food and labour, and now for land and resource grab. Moreover, the term 'underdevelopment' must not be seen only by the wealthy and powerful First World's standpoint. The villages and these communities were stable, sustained and dependent on each other but not on the uncertain, volatile market world where without cash, a man is crippled. However, since they don't comply with the mainstream definition of the developed societies, it does not mean that they are.

The lens of political ecology (Blaikie 2008; Robbins 2012) is a holistic approach that does not isolate but looks at the intertwined nature of social, political and economic policies which eventually cause environmental change. It helps revealing urbanisation as a multifaceted process of capitalist accumulation which is not limited to population shift and infrastructure building in the city centres to make them engines of growth. Urbanisation as development policy in GS (Clarke 2018) is imposed to serve the ends of global capitalist expansion. As a matter of fact, from the 1980s and early 1990s, capitalism strategically shifted to neoliberal policies and revolved into phase called neoimperialism (Enfu, and Global, L. B. H. I. M. P 2021). Critical development studies and the political ecology (Bobby Banerjee et al. 2023; Watts 2017) literature illuminates the web of actors, institutions, political dimensions and the power dynamics of that shape, state policies and the role of 'money' in expelling, displacing, alienating populations, jeopardising ecologies and the emergence of resistance against extraction from local political ecology by the indigenous communities. Capitalism in its neoimperialist incarnation is unapologetically sabotaging the stable environments and creating sacrificial zones. This stream of critical literature has predominantly focused on case studies from South Asia, East Asia, Latin America or GS countries.

Rapid urbanisation across the Global South is resulting in the large-scale commodification of peri-urban and rural land under a neoliberal paradigm of capitalist development (Banerjee-Guha 2009; Murray 2004; Shin 2016; Smit 2021) and has created free markets and promoted privatisation resulting in multi-faceted destabilised ecologies. The most talked and raised issues are displacement, dispossessions, loss of livelihoods, plundering of natural resources, food insecurity, waste, environmental and ecological degradation (Bebbington and Batterbury 2001; Hall 2013; Narain 2009; Neef et al. 2013; Sali and Sali 2012; Simon 2008; Zoomers 2010). The process begins by acquiring land and everything else follows.

3.3 Theoretical Framework for the Book

> *Underdevelopment is not due to the survival of archaic institutions and the existence of capital shortage in regions that have remained isolated from the stream of world history. On the contrary, underdevelopment was and still is generated by the very same historical process which also generated economic development: the development of capitalism itself.* (Frank 1966).

Andre Gunder Frank introduced this theory in 1966, and his work on 'Capitalism and development in Latin America' was considered to be one of the most crucial works on dependency theory. He argues that underdeveloped countries are not to be blamed for the failure; in fact, in globalised and highly interconnected world, the Western nations purposely failed to develop countries considered to be lagging in the race economic progress. Further, he says that core nations such as the USA and UK are the elite metropolis, and they happen to progress on the shoulders of the peripheral nations by treating them merely as a source of cheap raw material and manpower. This sort of exploitative relationship can be seen historically via slavery and colonisation in the past and in current times through international trade, the rise of multinationals (and financial institutions) and the reliance of undeveloped countries on Western or foreign aid (Ahmed 1975). Frank also draws a difference between underdeveloped and undeveloped by saying that "all developed countries were assumed to go through the same path of development and rest of the world needed to catchup" is a false notion. He also criticises modernisation theory of Walt and Rostow which states that countries need to pass through a linear process or stages from traditional to modern by acquiring practices and values from the developed world. He made a point by referring that such development theories are failed to see that how capitalist system was a cause of underdevelopment and not a solution. This whole chain of metropolis and satellite can even be seen within a country as well.

Although critiques have been levelled against Frank's theory (Amin 1976; PSIR 2021), many argue that it presents an overly simplistic and generalised representation of underdevelopment. The theory foregrounds economic factors while giving limited attention to political, social, cultural, and environmental dimensions that also contribute significantly to the processes of underdevelopment. The other criticism is that Frank puts blame on developed countries for the other half underdeveloped through exploitation or vertical relationship between metropolis and peripheral countries, but it does not include the historical ties due to which the dependency structure was built in the first place. For instance, what were the colonial reasons that gave rise to structural dependency or why India could not have scientific development but eventually got dependent on the western countries. Also, the other critique to dependence theories is that it is static in nature, and it draws a conclusion that peripheral countries will forever stay in the state of underdevelopment due to exploitation from the core and they will never progress. Whereas the practical example that negated this notion is the emergence of tiger economies in the late 1970s and 1980s, which were export-led countries such South Korea, Taiwan, Hong Kong and Singapore.

Despite coming across the criticism and after consulting different theories, I have realised that the book most appropriately falls under Frank's (1966) theory 'the development of the underdevelopment'. The book majorly argues that when space and place go through transformation, it is not just a visible change that can be seen through naked eye, but the invisible changes directly associated with intangible assets peculiar to ecologies and communities need to be deeply explored to see the impact of development on the process of change. In this particular study, the dependence of main city on the periphery was initially in terms of getting the produce, but the neoliberal policies for the economic expansion have commodified agricultural lands, agrarian landscape and livelihood which is not just a vocation, but a lifestyle composed of traditions, culture, norms and values. This system once provided people not only with food but also with irreplaceable assets such as security, safety, togetherness, identity, mutual care, and the social networks through which collective power was exercised. These forms of wellbeing emerged from the local ecology and could never be purchased on the market. With the acceleration of resource grabbing, communities have experienced a profound deterioration in the quality of land-based life. The evidence makes clear that money, and the masking of landscapes with imported symbols of modernity, does not elevate these communities. Instead, it dismantles the social and ecological worlds that once sustained them. These changes are driven by the desire to build posh gated communities and fashion a 'world-class' city modeled on Western ideals of modern living, aimed at attracting investment at the expense of already marginalised populations. The villages are not only undergoing underdevelopment through the erosion of agrarian structures but also suffering the loss of intangible forms of life that once enabled meaning, dignity, and everyday wellbeing. In analysing these dynamics, this work engages Frank's framework while also foregrounding the cultural, political, social, and economic dimensions that shape contemporary processes of dispossession.

3.4 Connected Themes

3.4.1 Land Question and Economic Models

Land acquisition is fundamental to many forms of accumulation, deemed 'development' with a promise of economic growth. Especially, since the period of (neo)-liberalisation, instead of state-led demand for land has not only been shifted but amplified to private demand for land for building infrastructure, industrial purposes and real-estate venture (Chandrasekhar and Ghosh 2002). World Bank, governments and their state-proposed polices favouring large-scale land acquisition (for development) cause impoverishment and loss of commons, jeopardising communities (particularly those living on state-owned land and have weak legal position) especially by denying their rights to traditional ways of living (Arrighi et al. 2010; Denis 2018a, b; Hall 2013; Kröger 2012) with aim of creating world-class cities (Goldman 2011) or 'megacity syndrome' (Mpungose and Maharaj 2022) through 'worldling from below' (Simone 2001). The multi-lateral donors like World Bank, big endorser of myopic, profit-centric, play selfish game of capitalism in the name of development through infrastructure building in Global South (Steel et al. 2017), rendering fringes as a *Tabula Rasa* (Kennedy and Sood 2016) or 'undeveloped land' as *Terra Nullius - the myth of empty land*, rendered as 'wasteland' during colonisation, which has legitimised local communities' displacement and dispossession (Bhandar 2018) which is still relevant today when one looks closely at the villages. These vulnerable geographies have not only given life to historical processes and standing as a proof of violent restructuring through homogenised city plans for the sake of corporations and imperial interests.

Levien (2017) explains the 'importance' of land and its dispossession for the booming development projects, especially since liberalisation. It is not just a response to fix the problem of over-accumulation, but it is highly political in character where state takes active part in resource distribution through extra-economic forces as well. He further argues that when coercion is persistently used to expropriate land from one group and transfer it to another, this constitutes a regime of dispossession. Such a regime becomes institutionalised through mechanisms designed to deprive rightful owners of their land. It has two defining features: first, the state's strategic withdrawal from protecting landowners in order to facilitate economic projects that benefit particular classes; and second, the creation of conditions that compel people to comply with requisition, producing obedience through both subtle pressure and overt force.

Hence, the development-based economic models liberated land markets, especially for the development of Special Economic Zones (SEZs), infrastructure building and urbanisation, and has accelerated the land grab (Zoomers 2010) from which benefits are not equally distributed but aimed at catering a certain faction at the expense of masses. Misra (2021) also foregrounds the unjust tied with social class and caste of those affected and its role in plating economic benefits and new ownership statuses up for the elite by exacting acquisition of farmlands from farmers in

SEZs in India. Despite their weak positions and the marginalisation of many due to hierarchies predicated on monetary power and caste, they put up resistance against state and corporates to safeguard traditional village economy.

3.4.2 *Peri-Urban Land: The New Hub for Investment*

The peri-urban region has borne the brunt of urban development in recent times. It is the site for land speculation nurturing real estate (Cartier 2011; Goldman 2011; Narain 2009), facilitating market in land across the countryside (Banerjee-Guha 2009; Chattopadhay et al. 2014; Levien 2012, 2013) and making periphery the landscape of change (Gururani 2020; Gururani and Dasgupta 2018). The rural villages on the periphery of the city are conveniently made into urban villages once it becomes the site for expropriation for urban development, and the involvement of state and urban plannings to encroach on farmlands to turn them into symbols of development makes the periphery 'spaces of suspension' which keeps settlers anxious and edgy (Ou 2021). The conversion of rural to urban landscape has been a lucrative business in real estate for the elites and has created new ways of grabbing mainly land and water through financialisation (Gutiérrez 2024). The land and soil acquisition processes in peri-urban villages explain the contradictions of capital. According to Harvey (2001), such processes behind building fixed space are indispensable for extracting profits and capital accumulation till a certain time which is then followed by demolition of the same landscape to make room for new 'spatial fix'. The following section explains how these processes are driving change in villages.

Peripheral urbanisation is a phenomenon based on reticular set of processes. It is not necessarily the city's expansion towards its periphery or margin, but it is a 'mode of production of urban space' that can happen anywhere. It involves the role of residents who actively take part in producing space, but state makes sure to clear the way for market in land to fuel capitalism which engenders displacement of 'autoconstructors' (local residents who renovate their houses and improve neighbourhoods at their own pace and time). However, the poor pay the price of upgraded localities as they become out of their reach. Hence, it is a process that always revolves around replacements, evictions and reproduction on a piece of land which is cheaper (Caldeira 2017). Such peripheral spaces are ideal for investment for several reasons, such as peripheries close to city provide flow of labour, capital, availability of infrastructure that offers easy transport, commute and trade, not under stern strictures of law which state developers and other actors involved can use in their advantage, and above all, land is cheaper (Gururani et al. 2021) for the sake of speculative urbanism (Goldman 2011).

The peripheries have not only revealed the real side of development led by urbanisation and discounts slogans of 'development goals' but also state's role imported economic policies and their pivotal place in destabilising ecologies, natural resources that must not treated anything less than 'national reserves', communities and how it has been siding the rich. For instance, according to Mishra and Narain (2018), the increased land value does benefit tiny section of well-heeled rural people, but there are other impacts such as declining clean drinking water quality, seepage problem, building of urban canals which begin with clearance of sandy land where vegetables used to be grown, expansion of wastewater irrigation, water theft, conflicts, loss of grazing lands and emergence of dengue fever in *Budhera* village. Dharia (2021) zeroes in on the corruption that is enmeshed as physical, political and economic phenomenon, which is culpable for incalculable social and environmental injustices. The irreparable consequences of state's support in gradual, haphazard turning of agricultural, forest land to encourage financialisation in land coupled with dispossession of many can be observed in multiple ways. The development of masterplan for the city expansion by encroaching into ecological-sensitive areas such as arable lands or forested turfs which is a huge attraction for real estate for the development of gated housing communities for money-making without giving due consideration to the provision of suitable basic services such as water, road networks, sewer and power supply system creates urban metabolic rifts.

3.4.3 Land Laws and Compensation Packages: Subtle Form of Violence

The land-use change based on land acquisition act brings in money via monetary compensations, an effective way not only to take over the land but setting up land market and converting farmers into small realtors and to camouflage social injustice via 'compensations' (Alias and Nasir Daud 2006; Ghatak and Ghosh 2011; Mahalingam and Vyas 2011; Mallik and Sen 2017; Sarkar 2014, 2015; Vijayabaskar 2010) and to bring ease to violators (Nair 2013) and engendering inegalitarianism (Tassadiq 2022). Especially in case of urban villages situated on the periphery, the negotiations over compensation between actors—residents, village committee, government officials and developer—could take years to settle (Nguyen 2017). However, young adults instead preferred to become realtors, speculators and rentiers through state compensations they received as they had no spot in urban high technology and capital-intensive developments (Zhan 2021). The land-use change is the harbinger of land dispossession for several rural dwellers. Once they go landless and become proletariats, many of them become fully 'proletarianised workers' (Chuang 2015; Swyngedouw and Ward 2020).

3.4.4 Land Interventions: Social in Character, Regardless of a Process

The urbanisation is a process of socio-ecological change, because land is not just physical terrain but extremely social; therefore, land question and land struggles are inevitable (Bose 2020). The urban expansion produces new environments and uneven geographies—new urban social and physical conditions—by undermining commons and local ecologies, where social actors defend and create own new environments with respect to class, ethnicity, gender conflicts and pronounced power struggles (Denis 2018a, b; Swyngedouw and Heynen 2003; Vij and Narain 2016), hence causing several long-term implications such as threat to food security, heat island affect and several other socio-economic problems (Du and Huang 2017; López et al. 2001; Satterthwaite et al. 2010; Wu 2008).

The peripheral urbanisation has also been called 'lust for land' (Denis 2018a, b), land grab/land rush (Neef et al. 2013; Neef and Singer 2015; Zoomers 2011), 'foreignisation of space' (Zoomers 2010), gentrification (Davidson and Lees 2010; Doshi 2015; Farahani 2013; Forrest 2016; Smith 1982, 2002; Waley 2016) and subaltern urbanisation (Mukhopadhyay et al. 2020). It could be a use of several terms to define the process, but it has been looked through a framework of accumulation by dispossession (Bhaduri 2015; Magdoff 2013; Shin 2016). He and Wang (2019) explain 'enclave urbanism' which in general is defined as homogenous unit that has its own social, cultural but economic features. However, such enclaves in recent times have several social implications due to symbolic representations attached with architecture—gates, walls, fences and boundaries that draw people who do not fit into this domain out. This spatial and social division notifies segregation and exclusion on the basis of class predominantly. Angotti (2012) connects urbanisation with orientalist logics to argue that development driven by high consumerism and individualized agency disproportionately benefits the affluent, who can readily secure land and other resources. This model overlooks and ultimately displaces the underprivileged communities who bear the costs of such development. Yet, as S. Li et al. (2012) observe, these exclusive enclaves can simultaneously generate high levels of social capital for their residents. Within these spaces, the affluent cultivate dense internal networks, shared norms, and forms of trust that further consolidate their privilege. What appears as 'community building' for some is premised on exclusion for others. The social capital that thrives inside gated or master-planned enclaves is thus produced through practices that restrict access, reinforce class boundaries, and spatially segregate those deemed outside the imagined modern, upwardly mobile urban subject.

Suburbanisation, as scholars have shown, is a process through which development models encroach on greenfield sites outside city cores, erasing villages and generating extensive urban sprawl (Gilbert & Gugler, 1992). This transformation is central to capital accumulation because, as Lefebvre et al., (1997) argues, the production of new urban space is a political-economic project. It aligns closely with Harvey's (2002, 2018) formulation of the "annihilation of space by time," whereby existing socio-ecological landscapes are rapidly overwritten to accelerate circulation, growth, and profit. Brenner and Schmid's (2015) work on planetary urbanisation further clarifies how these processes extend urban logics far beyond the city, while Shatkin (2016) highlights how suburban mega-developments in South and Southeast Asia commodify peripheral land to serve speculative and elite-driven growth. Together, these perspectives illustrate how suburbanisation depends on the systematic restructuring and destruction of rural and agrarian spaces, transforming them into frontiers for capital expansion. The development of suburbs is intended to centralise capital to urban areas by reducing entire society to urban society (Smith 1982), a sphere where there are series of urban and environmental processes that negatively affect some social groups and benefit others (Swyngedouw and Heynen 2003). The free market, liberalisation in land market, commodification and other moves for capital accumulation have been doing cultural damage and threatening the moral economy (Scott 1977) of close-knit communities.

3.4.5 Gentrification

Ruth Glass introduced the concept of gentrification in 1964 in city, and later, it was extended to rural areas in the 1990s (Lu et al. 2022). Gentrification is another urban process which assures social that carries the legacy of capital mode of production. Smith (1982) explains the process as the replacement of working-class residential areas by middle-class (higher social class) of people who have a better buying capacity such as landlords or developers; however, he also makes a distinction between gentrification and urban redevelopment, i.e. the construction of new infrastructure on land which is already developed but doesn't involve rehabilitation as gentrification does. Whereas Tomba (2017) also explains gentrification as replacement of one class by another which is higher in social hierarchy, but it is more than that. Further, he adds, it is not just a phenomenon to increase value of land and reshuffling of different aspects of a city such as political, economic and cultural, rather it is significant for the state to ascertain its ownership and control over land. Shin (2016) sheds lights on pivotal role of gentrification and redevelopment in freeing real-estate properties to attract investments and causes displacement in order to fuel speculative urban accumulation. Hence, the defining characteristics of gentrification in urbanising cities involve class replacement and social reshuffling, which produce multiple forms of displacement. These include not only physical eviction but also what Atkinson (2015) terms symbolic displacement, where long-term residents experience alienation and a loss of belonging as their environments are remade for new, affluent users. In many South Asian contexts, scholars observe similar dynamics: Ghertner (2015) shows how aesthetic norms of "world-class" urbanism delegitimise informal or agrarian settlements, while Shatkin (2008) demonstrates how land restructuring feeds speculative real-estate regimes across the region. State intervention in land markets, including the transfer of land to wealthier groups, reinforces uneven development and accelerates capital expansion, a process that aligns with Marx's idea of the annihilation of time and space. Smith (2002) further links gentrification to globalisation through the rise of 'global cities' that are reconfigured to attract foreign direct investment. In this context, neoliberal urbanism produces interlinked socio-economic and spatial transformations that actively reorganise urban and peri-urban frontiers, privileging elite mobility while deepening dispossession for marginalised communities.

Finally, high real-estate prices expel local residents. All the processes happening on land (Li 2014) explain that land which is often treated as a thing or a commodity is not piece of rug but a whole ecology; its physical entity provides resources that further forge relation with communities and surroundings which give 'land' a social character too. Similarly, Yang et al. (2018) point out rural gentrification (expansion from city to outskirts, a process that involves revamping of houses and public areas) as urban phenomenon that has transformed villages in China both socially and spatially, and this change has been led by predominantly villagers based on landownership. It does benefit landlords, but due to increasing rent gaps and property prices, young people have to move out. Author pinpoints the role of financial gains through rent-seeking which drives gentrification. Lu et al. (2022) write about the impact of rural gentrification on one of the villages in China and how physical and social landscape changed with in-migration of the new emerging middle class. Three prominent differences could be seen in terms of production, consumption and the role of government in the process.

3.4.6 Social Capital and Community Spirit

Social capital is typically defined as set of shared values, traditions and networks that forges an idea of togetherness amongst community members due to reciprocity and interdependence which act as a safety blanket. Woolcock and Narayan (2000) used maxim, 'It's not what you know, it's who you know', which almost sums up the concept of community building. Scott (1977) found the culture of sharing, working together, having common land and generosity towards each other as a basis for their resilience against tough times by maintaining subsistence. Marglin (2008, 2010) underscores the role of market economy in breaking down the communities. Deep social ties that bind people together and form community are constantly getting weakened by dominant market-driven economic ideology that has generated chaotic, individualistic and imbalanced human life. He uses the example of Amish people who refused to kneel against the market world. Similarly, Velayutham and Wise (2005) discussed a village of Tamil Nadu which they call 'translocal village' to explain how caste and kin-based ties forming a community provide a certain social control, fraught with responsibilities, morals, trust, networks and obligations that help maintain the moral economy (Scott 1977) of a village even from afar.

James C. Scott's later work, *Weapons of the Weak* (1985), extends this argument by demonstrating how communities under pressure from market forces and state encroachment often resist through subtle, everyday practices rather than overt rebellion. Foot-dragging, silence, selective compliance, gossip and the invocation of past moral orders operate as quiet repertoires of resistance. In Lahore's peri-urban frontier, these practices surface in the guarded negotiations villagers undertake with developers, in their reluctance to release land too hastily, and in their nostalgic retelling of agrarian pasts as moral critique of the present. These forms of everyday resistance do not halt dispossession, but they reveal the persistence of ethical benchmarks of justice and reciprocity, against which the inequities of neoliberal urbanism are measured.

Such perspectives remind us that social capital in peri-urban Lahore is not merely a static inheritance but an arena of struggle. Polanyi (2001) mentioned commodification for the societal collapse especially culturally, as 'value' is seen and measured differently in market world which is a flip side of pre-capitalist societies' perception of intangibles—invisible to the naked eye yet as fundamental and requisite to human life as food and water. Communities are being deprived of essential yet often unseen determinants of quality of life, such as trust and reciprocity, which become frayed, reconfigured, or unevenly stretched under urban pressure. At the same time, these social relations are strategically mobilised in ways that remind us that villagers are not passive recipients of urban takeover. Rather, they remain active participants who negotiate, resist, and shape the very processes through which urbanisation unfolds. Now, 'value' lies in the capitalist mode of production and is equated with the 'amount' of labour being used to make a 'commodity', also known as labour productivity; a productivity, commodification or accumulation carries a character of plunder and appropriation (Moore 2015). The frontiers play a significant role for capitalism to thrive and to strategise transformation by cheapening work, nature, lives, services, food and care not only by establishing economic power but also by exercising control over a 'web of life'. This has devastated the ecologies, indigenous communities and society–nature relation to keep the capitalism going (Patel and Moore 2019). Social capital, then, operates both as memory of what has been lost and as a weapon—however quiet—in on-going negotiations over land, life and livelihood. The conservation and preservation are nothing more than a 'bourgeois environmentalism' (Baviskar 2018; Crowley 2022), whereas the environmental protection is only possible if communities are left in charge. Sinha et al. (2013) help understanding the relation between local communities and resources and the importance of governance in the time of encroachment and heartless privatisation. At the same time, it allows to fathom the concept of sustainability and the role of communities in protecting these resources. Vandana Shiva's chapter in the book called 'Governance of Commons and Livelihood Security' (Shiva 2013) explains that the local communities are also the guardian of their resources as they are fully or partially dependent on commons for livelihood and sustenance. It is one of the characters of the sustained ecologies that they view earth as its mother and enjoy nature's bounties and in return maintain ecological-centric lifestyle. The biodiversity, water, soil, air, forests, etc. are of utmost importance to the locals because it provides them freedom and self-subsistence. On the other hand, as Locke explains that the capitalism thrives on the freedom to 'steal'. Hence, enclosures, privatisation of resources, rendering ecologically sensitive areas such as forests as 'wastelands' by the British and introduction of state policies to keep local communities astray by denying their right to access, is the cause of impoverishment, whereas enclosures as 'development' is the 'progress' for the corporations, privateers or the colonisers. However, Garrett Hardin's theory of 'Tragedy of the Commons' says otherwise (Hardin 1968).

In traditional rural culture, 'value' lies in character and exchange of intangibles such as trust, conviction, networks and barters not only in terms of grains and food but also with regards to tough times, emotions and moments of grief and happiness. Having said that, there is body of literature as critique of 'trust' and 'networks'. As Levien (2015) discussed two famous theories on social capital to fathom social change brought by economic processes in rural Rajasthan, Bourdieu's individual social capital which works well in unequal societies could be a huge resistance to inclusive development. The other one is Robert Putnam's theory of collective social capital according to which norms, trust and networks benefit not only individuals but higher the social capital, higher will be mutual benefits through collective action. However, author says that individuals with strong network or connection, more so due to their hierarchal position or contacts with influential people, get benefitted the most through such state-led development. In fact, the absence of collective social capital which has been eroded either due to exploitation in the hands of those whom people trusted or headmen who were more concerned about their own gains has weakened the resistance. As a matter of fact, the contemporary economic processes favour the individuals with strong networks; otherwise, it would have become an obstacle to development. Moreover, the migration trend in and out of the villages also changes the dynamics. Huang et al. (2022) found that social capital has declined not only in its structural form, which concerns networks and interactions among actors, but also in its cognitive dimensions, such as shared norms, trust, and mutual understanding. The stock of conviviality, identity, and the sense of being known has long been indispensable for resolving conflicts and maintaining everyday peace, and its erosion has significant implications for community cohesion. Nonetheless, Edwards (2004) shares a feminine perspective on social capital where family is the basic social institution and how the fundamental unit called 'family' and social capital, considered to be the basis of social cohesion, are strong but weak and frail at the same time that run on gender division, power relations and intergenerational relationships.

Eglar and Chowdhry (2010) provide an extensive account of the Punjabi village of *Mohla*, offering valuable insight into how rural settlements are structured and organised through caste and class hierarchies, the rural economy, and systems of exchange and reciprocity articulated through the concept of *"vartan bhanji"*. Their work illustrates the significance of land not only as a source of livelihood but also as a marker of prestige and a multidimensional asset that binds generations together and helps sustain *biraderi* (brotherhood) networks. Punjabi villages called *Mohla*. However, the place also went through a transformation in relation and co-dependency through somewhat disappearance of *vartan bhanji* when a village shifted from rural to market economy and influenced by the symbols of modernity. Irrespective of the onset of modernity be it from urbanisation, infrastructure building, industrialisation, highways and education to reproduce labour to increase the supply of workers for the factory owner and corporation locally and internationally, the consequences have commonality. This can be pinpointed through studies recording transformation in creating new place value and social change associated with it. Studies from across Pakistan reveal similar patterns of transformation. Essa (2016) documents the impact of tourism and the Karakoram Highway (KKH) in Hunza, showing how rapid infrastructural and economic change alters local livelihoods and social relations. Likewise, Batool et al. (2016), in their study of Bhera city in central Punjab, trace declines in health, water and food quality, the dismantling of small local businesses, and shifts toward modern education and urban employment that contribute to new forms of diaspora. These processes generate rising stress, deepening inequality, and the erosion of cultural and traditional values. The spread of gadgets, material consumption, and other symbols of capitalist 'progress' are often celebrated as indicators of development, yet they deplete and destabilise societies that were once rich in their own ways. Their forms of wealth did not align with mainstream metrics of prosperity, but they represent alternative modalities of affluence that are now being steadily undermined.

The studies on communities not only establish the impact of 'development' but also unfold reasons for the present situation in the world, i.e. chaos, violence, neoliberal and neocolonial, neoslavery, distrust, depression, diseases, broken family structure and list never ends. However, the places like villages, forests, inhabiting marginalised communities and peasantry give a wakeup call to find solutions to save of what is left (if any). The agrarian–urban frontier (Cowan 2024; Khan 2024) is studied in different ways. This transformation has revived the agrarian question in both its classic and contemporary formulations. Scholars from Kautsky et al., (1988) to Bernstein (1993) have long emphasised how capitalist expansion reshapes social relations around land, labour, and production. In the South Asian context, these debates illuminate how peasantry is not merely an economic category but a lived lifestyle grounded in an intimate alliance with nature, which historically secured subsistence and ecological sustainability. As land becomes increasingly commodified, this way of life is destabilised, signalling a deeper shift in the agrarian structure. Political ecologists further highlight that the very landscapes once embedded in systems of reciprocity and ecological care are being absorbed into what Searle (2016) describes as a "landscape of accumulation," where land is reconfigured primarily as a financial asset rather than a basis for social and ecological reproduction. One of the takeaways from the peasantry is that society and nature were not competitors, and people were ecologically conscious and aware of their limits and also to respect nature's limits too. The aim is completely opposite in capitalism, where greed is the only religion, and necessity is replaced with urge of having more (Coman 2024).

3.4.7 Women and their World

According to Agarwal (1989), the women especially from agrarian and tribal settings were considered to be oppressed, overburdened and marginalised due to various reasons such as access to basic necessities, property rights, access to land that goes to male members, varied earnings' opportunities and work environment (which treats workers according to their gender) and dismissal of ownership to the means of production, which keep them submissive to men. Despite all this, women have a deeper connection with their surroundings be it forceful or imposed gender roles, but their relation with ecology makes them integral part for safeguarding resources, and they have played a pivotal role in various environmental conservation movements. Further, she mentions the holistic approach of women towards the ecology when they put up resistance in *Chipko* movement. The women did not look at trees or forests only as the source for providing resources, but they sensed hazards that would happen with felling of trees, as they have more direct contact with natural resources (Mago and Gunwal 2019). However, when it comes to gender-based despotism, radical feminism finds patriarchy as a primary reason for women subordination and inequality. However, Comanne (2021) says that gender-based inequality works for capitalism. Nancy Fraser's interview for the book called *Feminism for the 99%* (Arruzza et al. 2019) explains the feminism as class issue. In order to resolve it, feminists need to be anti-capitalist; otherwise, they will keep empowering a certain elite faction and leave the masses to exploitation and abuse. It underscores that women's rights and gender equality cannot be understood and resolved unless it must be seen through class question which is created by capitalist mode of production (Malatesta 2019). The globalisation has also developed a 'global care chain', and now developing countries are experiencing not only brain drain but also care drain also, where women are major contributors who work in developed countries as nannies or domestic help. The creation of 'invisible ecology of care' or the flow of care is not only from the poor to rich counties, but even within a country, it is visible to see rural and urban areas within the same city. Love and care are renewable resources, a 'new gold' which is in demand for extraction from the GS by the GN (Hochschild 2004). The intangible assets, their commodification and the trap of market economy are stripping traditional societies and marginalised communities and not only depriving mothers from their traditional role but also depleting children of mother's love. Farris (2013) highlights that neoliberalisation has an immense impact on the commodification of care. Similarly, Smith (2008) explains the role of state in dividing society into different classes which it regulates to serve the interests of the upper or ruling classes. Moreover, state is also responsible for causing women's oppression through gender-based division of labour. The migration of male members to cities to serve factories leaving females behind in the village has made women free, independent and empowered when it comes to decision-making but has increased (even if it is for short term) the workload and responsibilities (Kaur 2020; Singh 2018). Women in rural areas are perceived as closer to the environment and to be fully immersed in ecological relations. It is easy to conclude that women and earth are inherently nurturers. Carpenter (2008) looks at the relation between women and livestock and women as primary collectors of fodder for the animals which itself is toilsome. Livestock husbandry and farming has a deeper connection in agrarian settings. The study is conducted in Pakistan which also draws comparison pertinent to labour division between pastoral and agricultural societies, owning of livestock and women as decision-makers in selling or buying animals and how to use their connection across the villages to find suitable persons. It also explores rainfed and irrigated farming systems and its impact on fodder which is a valuable resource to feed livestock especially for the poor households. For instance, the shrinkage of uncultivated land and wastelands led to fodder reduction. This in turns underscores the importance of communal land and its protection.

Akhtar et al. (2019) take southern Punjab as a case study to explain role of women in socio-economic sphere. The study explains the integral role women play by taking care of several chores such as taking care of families, livestock and sowing till harvesting on agricultural fields. They are active in economic activities to help their men to generate money, yet their contributions are not acknowledged and valued. Women have almost no place in decision-making at family and community levels. They have taken *Dera Ismail Khan* as a study area to analyse the rural women in decision-making when it comes to income generation for family, such as agricultural patterns, land use, crop selection and also the minimal-to-none say in family-related decisions. The study found out that the women do have better understanding towards budgeting the finances, yet their part in making decisions such as in education, marriage, family resources or income is not much appreciated and welcomed. Petesch and Badstue (2020) took 32 villages from South Asia including areas from *Khyber Pakhtunkhwa* (KPK), Pakistan, to look at the relation between gender norms and poverty (reduction). They acknowledge the active involvement of women in agriculture and other economic activities to bring wellbeing to their families. However, it does not change much when it comes to men's domination over women and resources. There were cases recorded by the study where women were the reasons for dragging their families out of poverty through small-scale agricultural enterprises, especially those who had a strong support system from men in their families (relations of domination), such as husband, father or a brother.

Feminism is now in its fourth wave, which began around 2012. While critics argue that the word 'wave' is misleading and does not do justice as presenting women struggle over time as oversimplified version of complex history. The core contribution of feminism and different schools such as radical, Marxist and liberal has raised several important issues such as role of class, race, security at workplaces, right to vote, property rights, unequal wages, subjugation, male dominance, sexual violence and patriarchal society that do not let women thrive. What is less explored is the question of religion and inclusion of women for whom the meaning of 'agency' or 'freedom' is not borrowed from the West, and their self-worth, aspirations, desires and lives have been shaped from nonliberal traditions (Mahmood 2006). The study is based on ethnographic account of women involved in piety movement in Cairo, Egypt, mosques. The study shows a perspective from women who have been regarded as oppressed and trapped in male authority. From the article, I would like to quote her writing related to 'veil' as women's voluntary decision as virtue of piety especially when it comes to a bodily conduct and idea of shyness—'*haya*' (that is synonymously used for someone being underconfident) and modesty.

3.4.8 Nostalgia

One of the aspects of this book is the question of nostalgia. Although most literature on nostalgia is found in psychology, this makes it challenging to incorporate the concept affectively within sociology, where emotions have historically received limited attention. Jacobsen (2023) notes that sociologists have tended to overlook nostalgia and often remain silent on the emotional dimensions of social experience. However, since the 1980s, sociology has increasingly recognised emotions as central to understanding social life and has begun to treat nostalgia not simply as an individual psychological state but as a socially embedded affect. Scholars of affect theory and emotional geographies, such as Liz Bondi (2005) and Ben Anderson (2014), emphasize that nostalgia emerges through relationships, memories, and attachments to place, and is deeply shaped by collective histories and shared practices. Memory studies further show that nostalgia plays a crucial role in how individuals and communities make sense of rupture, loss, and displacement, particularly in contexts of rapid urbanisation. Integrating nostalgia through this sociological lens allows us to see it as a relational and meaning-making force, illuminating how people negotiate change while holding on to affective anchors that provide continuity, identity, and a sense of belonging.

Nostalgia is somewhat defined as the longing for the past, a desire to reverse the time or to bring the former phase of life again which is cherished profoundly. It can be individual or collective linkage to the past but is a real social emotion (Juhl and Biskas 2023). Nostalgia is usually referred as 'bittersweet' that on the one hand recalls the hurtful experience of loss (of good times), while, on the other hand, the 'sweetness' part is associated with warmth and tenderness of the memories from the past (Sedikides and Wildschut 2016). Nostalgia is also viewed as a combination of memory, idealisation and romanticisation of the past. It has the potential to diminish the negative and glorify the positive experiences of the past life (Davis 1979). It is real because it acknowledges what an individual once had. There can be several aspects pertinent to loss, but the additional dimension to the 'nostalgia as loss experience' is 'the sense of loss of individual freedom and autonomy with the disappearance of genuine social relationships' (Turner 1987). The other relevant aspect of nostalgia in the book is called 'solastalgia', which means that people or communities that experience connectedness to their environment feel afflicted and distraught due to the environmental change (Albrecht et al. 2007). Nostalgia is also a source of identity especially when one's sense of belonging to a place or community is threatened or sabotaged due to any reason, and this is when nostalgia awakens by attaching one's strings of identity in present times to the previous one and provides cushion for the self-continuity (Sedikides et al. 2015). Literature also differentiates between private and collective nostalgia (Wildschut and Sedikides 2023) by distinguishing between the former and 'historical nostalgia'—a collection of remuneration and celebration of the past (Stern 1992). The reciprocity and exchange especially of intangibles develop a community. The sense of identity and affinity is of paramount significance; therefore, social connectedness is not just a basic psychological need but also imperative for wellbeing (Baumeister and Leary 1995). Nostalgia in social context is connected to shared set of values, ideologies and circumstances. Therefore, it needs to be given more attention and value rather than sweeping it under the rug by calling it just the 'longing for old days'. Even if it is, it carries substantial information to answer *why* through visiting personal journal of an individual or a community's life events and a journey of transformation. Nostalgia is the invisible documentation of physical, emotional and cognitive landscapes.

3.5 Different Outlooks Towards the Process

Pakistan shares historical conditions which have shaped the trajectory of the nature of urban development mentioned above. Historical accounts of economic change in Lahore (Chattha 2012), rural urbanisation and agriculture (Mughal 2019), the changing patterns in land governance and marginalisation of farmers on state-owned land in *Okara* (especially military as a primary actor), the role of CPEC[3] on rural people (Mehmood 2019) and the power of state in Pakistan (Akhtar 2006) are available. Qadeer (1996) has written an extensive account on unfolding the urban policies in Pakistan. State policies play a similar role in other metropolitan cities, as seen in Karachi's periphery, where land grabs justified in the language of development and infrastructure have enabled the transfer of vast tracts of land to powerful actors, accelerating dispossession and reshaping agrarian and informal settlements into speculative urban frontiers. Studies also exist related to the villages (*Baloch goths*) that have been thrown into disarray with huge development-induced displacement (Anwar 2012, 2014, 2018; Hasan 2015). There are studies on conversion of arable land to construction sites and its impacts on the resources (Zaman and Baloch 2011), whereas Anjum and Hameed (2007) wrote in favour of peripheral housing by taking systemic approach at the end of policy makers and also put emphasis on better transportation facilities. Cermeño (2021) highlighted planning and policies in land transformation on the Lahore's periphery. However, there is no holistic study of social and ecological change in peri-urban villages of Lahore built through an extensive collection of narratives especially from females and the lower classes. They have experienced rapid in situ urbanisation with growth of more capitalist developmental ventures. This has led to cash invasion in the communities which were cash poor historically. This study shows how peri-urban marginalised

[3] China-Pakistan Economic Corridor

and the poor communities have been living through development-led urbanisation with the help of ethnographic research to document gain and loss account. Ethnographic research itself in Pakistan has been fairly limited and remains under-theorised, given that most of the ethnographic theories have come from Western scholars studying non-western cultures (Ugwu 2017).

References

Agarwal B (1989) Rural women, poverty and natural resources: sustenance, sustainability and struggle for change. Econ Polit Wkly 24(43):WS46–WS65

Ahmed F (1975) Aiding Underdevelopment in Pakistan. MERIP Reports 42:19. https://doi.org/10.2307/3693727

Akhtar AS (2006) The state as landlord in Pakistani Punjab: peasant struggles on the Okara military farms. J Peasant Stud 33(3):479–501. https://doi.org/10.1080/03066150601063058

Akhtar S, Ahmad S, Yousaf H (2019) Role of women in social and agricultural decision-making in the rural areas of the southern Punjab (Pakistan). Int J Humanit Soc Sci 5. https://doi.org/10.14445/23942703/IJHSS-V5I5P106

Albrecht G, Sartore G-M, Connor L, Higginbotham N, Freeman S, Kelly B, Stain H, Tonna A, Pollard G (2007) Solastalgia: the distress caused by environmental change. Australas Psychiatry.: Bulletin of Royal Australian and New Zealand College of Psychiatrists 15(Suppl 1):S95–S98. https://doi.org/10.1080/10398560701701288

Alias A, Nasir Daud MD (2006) Payment of adequate compensation for land acquisition in Malaysia. Pac Rim Prop Res J 12(3):326–349. https://doi.org/10.1080/14445921.2006.11104213

Ameel L, Gurr JM, Buchenau B (2023) Narrative in urban planning: a practical field guide. transcript Verlag. https://doi.org/10.1515/9783839466179

Amin S (1976) Unequal development: an essay on the social formations of peripheral capitalism. Harvester Press

Anderson B (2014) Encountering affect: Capacities, apparatuses, spaces. Ashgate Publishing, Ltd

Angelo H, Wachsmuth D (2015) Urbanizing urban political ecology: a critique of methodological Cityism: urbanizing urban political ecology. Int J Urban Reg Res 39(1):16–27. https://doi.org/10.1111/1468-2427.12105

Angotti T (2012) The new century of the metropolis (0 ed.). Routledge. https://doi.org/10.4324/9780203114193

Anjum GA, Hameed R (2007) The dynamics of colonization of peripheral housing schemes and policy options in case of Lahore. Pak J Eng Appl Sci. https://journal.uet.edu.pk/ojs_old/index.php/pjeas/article/view/214

Anwar N (2012) State power, civic participation and the urban frontier: the politics of the commons in Karachi. Antipode 44(3):601–620. https://doi.org/10.1111/j.1467-8330.2011.00920.x

Anwar NH (2014) Urban transformations: brokers, collaborative governance and community building in Karachi's periphery. South Asian Hist Cult 5(1):75–92. https://doi.org/10.1080/19472498.2013.863011

Anwar NH (2018) Receding rurality, booming periphery 53(12). https://www.epw.in/journal/2018/12/review-urban-affairs/receding-rurality-booming-periphery.html

Arrighi G, Aschoff N, Scully B (2010) Accumulation by dispossession and its limits: the southern Africa paradigm revisited. Stud Comp Int Dev 45(4):410–438. https://doi.org/10.1007/s12116-010-9075-7

Arruzza C, Bhattacharya T, Fraser N (2019) Feminism for the 99 percent: a manifesto. Verso

Atkinson R (2015) Losing one's place: narratives of Neighbourhood change, market injustice and symbolic displacement. Hous Theory Soc 32(4):373–388. https://doi.org/10.1080/14036096.2015.1053980

Banerjee-Guha S (2009) Neoliberalising the "urban": new geographies of power and injustice in Indian cities. https://www.epw.in/journal/2009/22/special-articles/neoliberalising-urban-new-geographies-power-and-injustice-indian
Batool F, Nadir R, Ghazanfar M (2016) Is modernity depleting Bhera *Vol.*6 No.1. https://lahore-journalofpolicystudies.org/issues.html
Baumeister RF, Leary MR (1995) The need to belong: desire for interpersonal attachments as a fundamental human motivation. Psychol Bull 117(3):497–529. https://doi.org/10.1037/0033-2909.117.3.497
Baviskar A (2018) Urban jungles: wilderness, parks and their publics in Delhi. Econ Polit Wkly. https://www.academia.edu/41666296/Urban_Jungles_Wilderness_Parks_and_Their_Publics_in_Delhi
Bebbington AJ, Batterbury SPJ (2001) Transnational livelihoods and landscapes: political ecologies of globalization. Ecumene 8(4):369–380
Bhaduri A (2015) A study in development by dispossession. Camb J Econ 42(1):19–31. https://doi.org/10.1093/cje/bex026
Bhandar B (2018) Colonial lives of property: law, land, and racial regimes of ownership. Duke University Press
Blaikie P (2008) Epilogue: towards a future for political ecology that works. Geoforum 39(2):765–772. https://doi.org/10.1016/j.geoforum.2007.07.004
Bobby Banerjee S, Maher R, Krämer R (2023) Resistance is fertile: toward a political ecology of translocal resistance. Organization 30(2):264–287. https://doi.org/10.1177/1350508421995742
Bondi L (2005) The place of emotions in research: from partitioning emotion and reason to the emotional dynamics of research relationships. In J. Davidson, L. Bondi, & M. Smith (Eds.), Emotional Geographies (pp. 231–246). Ashgate Publishing
Bose S (2020) Attachment to place and territoriality in Nandigram land struggle, India. Hum Geogr 13(2):139–153. https://doi.org/10.1177/1942778620944383
Brenner N (2014) Implosions/explosions: Towards a Study of Planetary Urbanization. Jovis
Brenner N, Schmid C (2015) Extending the theory of planetary urbanization: Towards a joint framework. Implosions/Explosions: Towards a Theory of Planetary Urbanization (pp. 154–173). Jovis
Bernstein E (with Tudor, H.) (1993) Bernstein: The Preconditions of Socialism (1st ed). Cambridge University Press.
Caldeira TP (2017) Peripheral urbanization: autoconstruction, transversal logics, and politics in cities of the global south. Environ Plann D Soc Space 35(1):3–20. https://doi.org/10.1177/0263775816658479
Carpenter C (2008) Women and livestock, fodder, and uncultivated land in Pakistan: a summary of role responsibilities. Soc Nat Resour 4(1):65–79. https://doi.org/10.1080/08941929109380743
Cartier C (2011) Urban growth, rescaling, and the spatial administrative hierarchy. Provincial China 3(1):Article 1. https://epress.lib.uts.edu.au/journals/index.php/provincial_china/article/view/2440
Celestina M (2016) "Displacement" before displacement: time, place and the case of rural Urabá. J Lat Am Stud 48(2):367–390. https://doi.org/10.1017/S0022216X15001212
Cermeño H (2021) Living and planning on the edge: unravelling conflict and claim-making in Peri-urban Lahore, Pakistan. Urban Planning 6(2):189–201. https://doi.org/10.17645/up.v6i2.3858
Chandrasekhar CP, Ghosh J (2002) The market that failed: neoliberal economic reforms in India. LeftWord Books
Chattha I (2012) Economic change and community relations in Lahore before partition. J Punjab Stud 19:193–213
Chattopadhay S, Dey P, & Michael J (2014) Dynamics and growth dichotomy of urban villages: Case study Delhi. Journal of Housing Science 38(2):81–94
Chuang J (2015) Urbanization through dispossession: survival and stratification in China's new townships. J Peasant Stud 42(2):275–294. https://doi.org/10.1080/03066150.2014.990446

Clarke M (2018) Global South: what does it mean and why use the term? https://onlineacademic-community.uvic.ca/globalsouthpolitics/2018/08/08/global-south-what-does-it-mean-and-whyuse-the-term/. Accessed Sep 2018

Coman J (2024) Patrick Joyce: 'The history of peasants is one of their silence or being silenced.' The Guardian. https://www.theguardian.com/books/2024/feb/03/patrick-joyce-remembering-peasants-a-personal-history-of-a-vanished-world-interview

Comanne D (2021) How patriarchy and capitalism combine to aggravate the oppression of women. cadtm. https://www.cadtm.org/How-Patriarchy-and-Capitalism-Combine-to-Aggravate-the-Oppression-of-Women

Cornea N (2019) Urban political ecology. In: Cornea N (ed) Geography. Oxford University Press. https://doi.org/10.1093/obo/9780199874002-0203

Cowan T (2024) Urbanization without guarantees: articulation, rentier capitalism and occupancy on the agrarian urban frontier. Urban Geogr:1–24. https://doi.org/10.1080/02723638.2024.2308399

Crowley T (2022) From bourgeois environmentalism to a just urban commons: despair and Hope in Amita Baviskar's Uncivil City. Conserv Soc 20(1):54. https://doi.org/10.4103/cs.cs_130_21

Davidson M, Lees L (2010) New-build gentrification: its histories, trajectories, and critical geographies. Popul Space Place 16(5):395–411. https://doi.org/10.1002/psp.584

Davis F (1979) Yearning for yesterday: a sociology of nostalgia (1. print.). New York : Free Pr. [u.a.]

Denis E (2018a) Urban desires and lust for land south. Métropolitiques https://shs.hal.science/halshs-01780595

Denis E (2018b) Urban desires and lust for land. The Commodification of Rural Spaces in the Global South

Dharia NV (2021) Embodied urbanisms: corruption and the ecologies of eating and excreting in India's real estate economies. Cult Anthropol 36(4):Article 4. https://doi.org/10.14506/ca36.4.11

Doshi S (2015) Rethinking gentrification in India: displacement, dispossession and the spectre of development. In: Lees L, Shin HB, López-Morales E (eds) Global gentrifications: uneven development and displacement. Policy Press. https://doi.org/10.1332/policypress/9781447313472.003.0006

Du X, Huang Z (2017) Ecological and environmental effects of land use change in rapid urbanization: the case of Hangzhou, China. Ecol Indic 81. https://doi.org/10.1016/j.ecolind.2017.05.040

Edwards R (2004) Present and absent in troubling ways: families and social capital debates*. Sociol Rev 52(1):1–21. https://doi.org/10.1111/j.1467-954X.2004.00439.x

Eglar ZS, Chowdhry FA (2010) A Punjabi village in Pakistan: perspectives on community, land, and economy. Oxford University Press

Enfu C, Global, L. B. H. I. M. P (2021) Monthly review | five characteristics of Neoimperialism. Mon Rev. https://monthlyreview.org/2021/05/01/five-characteristics-of-neoimperialism/

Essa S (2016) Khas Daes Kay Aam Log place and people of Karimabad, Hunza. Lahore Journal of Policy Studies 6(1) https://lahorejournalofpolicystudies.org/issues.html

Farahani I (2013) Vanished in gaps, vanquished in rifts: the social ecology of urban spatial change in a working class residential area, Peykan-Shahr, Tehran, Iran. Journal of Political Ecology 20:395–412

Farris SR (2013) Neoliberalism, migrant women, and the commodification of care. Sch Fem Online:11.1–11.2. https://sfonline.barnard.edu/neoliberalism-migrant-women-and-the-commodification-of-care/

Forrest R (2016) Commentary: variegated gentrification? Urban Stud 53(3):609–614. https://doi.org/10.1177/0042098015616308

Frank AG (1972) The development of underdevelopment. In J. D. Cockcroft, A. G. Frank, & D. L. Johnson (Eds.), Dependence and underdevelopment: Latin America's political economy (pp. 3-17). Anchor Books

Frank AG (1966) The development of underdevelopment. Mon Rev:17–31. https://doi.org/10.14452/MR-018-04-1966-08_3
Ghatak M, Ghosh P (2011) The land acquisition bill: A critique and a proposal. Economic and Political Weekly 46(41):65–72
Ghertner DA (2015) Rule by aesthetics: World-class city making in Delhi. Oxford University Press.
Gilbert A, Gugler J (1992) Cities, poverty and development: Urbanization in the Third World (2nd ed.). Oxford University Press
Ghose R (2004) Big sky or big sprawl? Rural gentrification and the changing cultural landscape of Missoula, Montana. Urban Geogr 25(6):528–549. https://doi.org/10.2747/0272-3638.25.6.528
Glover W (2021) The other agrarian urbanisation: urbanism in the village. Urbanisation 6(1):35–48. https://doi.org/10.1177/24557471211016591
Goldman M (2011) Speculative urbanism and the making of the next World City. Int J Urban Reg Res 35(3):555–581. https://doi.org/10.1111/j.1468-2427.2010.01001.x
Gururani S (2020) Cities in a world of villages: agrarian urbanism and the making of India's urbanizing frontiers. Urban Geogr 41(7):971–989. https://doi.org/10.1080/02723638.2019.1670569
Gururani S, Dasgupta R (2018) Frontier urbanism: urbanisation beyond cities in South Asia. Econ Polit Wkly 53:41–45
Gururani S, Kennedy L, Sood A (2021) Engaging the urban from the periphery. South Asia Multidiscip Acad J. https://www.academia.edu/48975311/Engaging_the_Urban_from_the_Periphery
Gutiérrez J (2024) Real estate oligarchs: elites and the urbanization of the land question in El Salvador. J Peasant Stud 51(2):489–511. https://doi.org/10.1080/03066150.2023.2252758
Hall D (2013) Primitive accumulation, accumulation by dispossession and the global land grab. Third World Q 34(9):1582–1604
Hardin G (1968) The tragedy of the commons: the population problem has no technical solution; it requires a fundamental extension in morality. Science 162(3859):1243–1248. https://doi.org/10.1126/science.162.3859.1243
Harvey D (2001) Globalization and the "Spatial Fix.". Geographische Revue : Zeitschrift Für Literatur Und Diskussion 3(2):23–30
Harvey D (2002) Spaces of capital: Towards a critical geography. Routledge
Harvey D (2018) The limits to capital. Verso books
Hasan A (2015) Value extraction from land and real estate in Karachi (pp. 181–198). https://doi.org/10.46692/9781447313496.011
He S, Wang K (2019) Enclave urbanism. In A. M. Orum, The Wiley Blackwell encyclopedia of urban and regional studies (1st ed., pp. 1–5). Wiley. https://doi.org/10.1002/9781118568446.eurs0089
Heynen N (2017) Urban political ecology. In: Richardson D, Castree N, Goodchild MF, Kobayashi A, Liu W, Marston RA (eds) International encyclopedia of geography, 1st edn. Wiley, pp 1–9. https://doi.org/10.1002/9781118786352.wbieg1110
Hochschild AR (2004) Love and gold. Nouv Quest Feminist 23
Huang Z, Liu Y, Wang Y, Jin S (2022) Does migration experience reduce villagers' social capital? Evidence from rural China. Appl Econ:1–13. https://doi.org/10.1080/00036846.2022.2115453
Jacobsen MH (2023) The sociology of nostalgia. Curr Opin Psychol 50:101556. https://doi.org/10.1016/j.copsyc.2023.101556
Juhl J, Biskas M (2023) Nostalgia: An impactful social emotion. Curr Opin Psychol 49:101545. https://doi.org/10.1016/j.copsyc.2022.101545
Kaur AP (2020) Impact of migration on gender roles: study of left behind wives in rural Punjab. Indian J Health Well Being 11(7–9):407–411
Kautsky K, Alavi H, Shanin T, Burgess P (1988) The agrarian question. Vol. 2. Zwan
Kennedy L, Sood A (2016) Greenfield development as tabula rasa. Rescaling, Speculation and Governance on India's Urban Frontier

Khan D (2024) Political economy of the 'agrarian–urban frontier' in Pakistan: agrarian transformation, social reproduction and exploitation. J Agrar Chang 24(3):e12582. https://doi.org/10.1111/joac.12582
Khan I (2020) Uneven development and the phenomena of housing societies in Pakistan (Urban Monograph Series 2020:3). Pakistan Institute of Development Economics.
Kröger M (2012) The expansion of industrial tree plantations and dispossession in Brazil. Dev Chang 43(4):947–973. https://doi.org/10.1111/j.1467-7660.2012.01787.x
Kugkhapan NT (2014) Narrative in urban theory. http://deepblue.lib.umich.edu/handle/2027.42/120397
Lefebvre H, Nicholson-Smith D, Lefebvre H, Lefebvre H (1997) The production of space (Reprinted). Blackwell
Levien M (2012) The land question: special economic zones and the political economy of dispossession in India. J Peasant Stud 39:933–969. https://doi.org/10.1080/03066150.2012.656268
Levien M (2013) The Politics of Dispossession: Theorizing India's "Land Wars.". Polit Soc 41(3):351–394. https://doi.org/10.1177/0032329213493751
Levien M (2015) Social capital as obstacle to development: brokering land, norms, and Trust in Rural India. World Dev 74:77–92. https://doi.org/10.1016/j.worlddev.2015.04.012
Levien M (2017) From primitive accumulation to regimes of dispossession. Oxford Scholarship Online. https://www.academia.edu/12684353/From_Primitive_Accumulation_to_Regimes_of_Dispossession_Six_Theses_on_Indias_Land_Question
Li TM (2014) What is land? Assembling a resource for global investment. Trans Inst Br Geogr 39(4):589–602. https://doi.org/10.1111/tran.12065
Li S, Zhu Y, Li L (2012) Neighborhood type, gatedness, and residential experiences in Chinese cities: a study of Guangzhou. Urban Geogr 33(2):237–255. https://doi.org/10.2747/0272-3638.33.2.237
López T d M, Aide TM, Thomlinson JR (2001) Urban expansion and the loss of prime agricultural lands in Puerto Rico. Ambio 30(1):49–54
Lu S, Rao X, Duan P (2022) The Rural Gentrification and Its Impacts in Traditional Villages—A Case Study of Xixinan Village, in China. Sustainability 14(16):10077. https://doi.org/10.3390/su141610077
Magdoff F (2013) Twenty-first-century land grabs: Accumulation by agricultural dispossession. Monthly Review 65(6):1–19
Mago P, Gunwal I (2019) Role of women in environment conservation (SSRN scholarly paper No. 3581051). https://doi.org/10.2139/ssrn.3581051
Mahalingam A, Vyas A (2011) Comparative evaluation of land acquisition and compensation processes across the world. Econ Polit Wkly 46(32):94–102
Mahmood S (2006) Feminist theory, agency, and the Liberatory subject: some reflections on the Islamic revival in Egypt. Etnografica 10:121–158. https://doi.org/10.33356/temenos.4633
Malatesta O (2019) A feminism aimed at liberating all women must be anti-capitalist [Interview]. https://jacobin.com/2019/10/nancy-fraser-feminism-anti-capitalist-99-percent-majority
Mallik C, Sen S (2017) Land acquisition policy and praxis: the case of peri-urban Delhi. In: Acharya SS, Sen S, Punia M, Reddy S (eds) Marginalization in globalizing Delhi: issues of land, livelihoods and health. Springer India, pp 115–126. https://doi.org/10.1007/978-81-322-3583-5_7
Marglin S (2008) Why thinking like an economist can be harmful to the community. Challenge 51(2):13–26. https://doi.org/10.2753/0577-5132510203
Marglin S (2010) The dismal science: how thinking like an economist undermines community (First Harvard University Press paperback edition). Harvard University Press
Mehmood A (2019) Postcolonial land governance in Pakistan: exclusionary practices on state-owned farms. South Asia Res 39(2):218–235. https://doi.org/10.1177/0262728019843708
Mishra P, Narain V (2018) Urban canals and peri-urban agrarian institutions—a study of Budhera village in Gurgaon. Econ Polit Wkly 53(37):51–58

Misra K (2021) Political domination and economic dispossession of farmers: the case of land Acquisition for Special Economic Zones in India. J Glob Dev 12(2):181–219. https://doi.org/10.1515/jgd-2020-0083

Moore JW (2015) Capitalism in the web of life: ecology and the accumulation of capital, 1st edn. Verso

Mpungose A, Maharaj B (2022) Megaprojects in the context of neoliberalism: socio-economic and spatial impacts of the proposed dig-out port in Durban, South Africa. Urban Forum 33(2):173–195. https://doi.org/10.1007/s12132-021-09442-8

Mughal MAZ (2019) Rural urbanization, land, and agriculture in Pakistan. Asian Geogr 36(1):81–91. https://doi.org/10.1080/10225706.2018.1476255

Mukhopadhyay P, Zérah M-H, Denis E (2020) Subaltern urbanization: Indian insights for urban theory. Int J Urban Reg Res 44(4):582–598. https://doi.org/10.1111/1468-2427.12917

Murray MJ (2004) The spatial dynamics of postmodern urbanism: social polarisation and fragmentation in Saão Paulo and Johannesburg1. J Contemp Afr Stud 22(2):139–164. https://doi.org/10.1080/cjca0258900042000229999

Nair J (2013) Is there an 'Indian' urbanism? 43–70. https://doi.org/10.5790/hongkong/9789888139767.003.0002

Narain V (2009) Growing city, shrinking hinterland: land acquisition, transition and conflict in peri-urban Gurgaon, India. Environ Urban 21(2):501–512. https://doi.org/10.1177/0956247809339660

Neef A, Singer J (2015) Development-induced displacement in Asia: conflicts, risks, and resilience. Dev Pract 25(5):601–611. https://doi.org/10.1080/09614524.2015.1052374

Neef A, Touch S, Chiengthong J (2013) The politics and ethics of land concessions in rural Cambodia. J Agric Environ Ethics 26(6):1085–1103. https://doi.org/10.1007/s10806-013-9446-y

Nguyen V (2017) Slow construction: alternative temporalities and tactics in the new landscape of China's urban development. City 21(5):650–662. https://doi.org/10.1080/13604813.2017.1375728

Ou T-C (2021) Spaces of suspension: construction, demolition, and extension in a Beijing migrant Neighbourhood. Pac Aff 94(2):251–264. https://doi.org/10.5509/2021942251

Patel R, Moore JW (2019) A history of the world in seven cheap things: a guide to capitalism, nature, and the future of the planet. University of California Press. https://doi.org/10.1525/9780520966376

Petesch P, Badstue L (2020) Gender norms and poverty dynamics in 32 villages of South Asia. Int J Community Well-Being 3(3):289–310. https://doi.org/10.1007/s42413-019-00047-5

Polanyi K (2001) The great transformation: the political and economic origins of our time (2nd Beacon Paperback ed). Beacon Press

PSIR (Director). (2021) Development of underdevelopment (Dependency Theory) by A.G. Frank [Youtube video]. https://www.youtube.com/watch?v=CSpHMKdxjPY

Qadeer MA (1996) An assessment of Pakistan's urban policies, 1947–1997. Pak Dev Rev 35(4):443–465

Robbins P (2012) Political ecology: a critical introduction, 2nd edn. Wiley

Sali G, Sali G (2012) Agricultural land consumption in developed countries. https://doi.org/10.22004/AG.ECON.126431

Sarkar S (2014) Speculation in land and the predicament of capitalist urbanization. SSRN Electron J. https://doi.org/10.2139/ssrn.2462653

Sarkar S (2015) Beyond dispossession. Comp Stud South Asia Afr Middle East 35(3):438–450. https://doi.org/10.1215/1089201X-3426289

Satterthwaite D, McGranahan G, Tacoli C (2010) Urbanization and its implications for food and farming. Philos Trans R Soc B Biol Sci 365(1554):2809–2820. https://doi.org/10.1098/rstb.2010.0136

Scott JC (1977) The moral economy of the peasant: rebellion and subsistence in Southeast Asia. Published in 1976. (Nachdr.). Yale Univ. Press

Scott JC (1985) Weapons of the weak: everyday forms of peasant resistance. Yale University Press

Searle LG (2016) Landscapes of accumulation: real estate and the neoliberal imagination in contemporary India. University of Chicago Press. https://doi.org/10.7208/chicago/9780226385235.001.0001
Sedikides C, Wildschut T (2016) Nostalgia: a bittersweet emotion that confers psychological health benefits. In: The Wiley handbook of positive clinical psychology. Wiley Blackwell, pp 125–136. https://doi.org/10.1002/9781118468197.ch9
Sedikides C, Wildschut T, Routledge C, Arndt J (2015) Nostalgia counteracts self-discontinuity and restores self-continuity. Eur J Soc Psychol 45(1):52–61. https://doi.org/10.1002/ejsp.2073
Shatkin G (2008) The City and the Bottom Line: Urban Megaprojects and the Privatization of Planning in Southeast Asia. Environment and Planning A: Economy and Space 40(2):383–401. https://doi.org/10.1068/a38439
Shatkin G (2016) Planning for mega-urban regions in South and Southeast Asia: A comparative analysis of institutional challenges. Planning Perspectives 31(3):433–456
Shin HB (2016) Economic transition and speculative urbanisation in China: gentrification versus dispossession. Urban Stud 53(3):471–489. https://doi.org/10.1177/0042098015597111
Shiva V (2013) Governance of commons and livelihood security. In: Sinha H, Kumar A, Xavier Institute of Social Service (Rānchī, India (eds) International conference on governance of commons and livelihood security, Ranchi, Jharkhand, India. Xavier Institute of Social Service
Simon D (2008) Urban environments: issues on the Peri-urban fringe. Annu Rev Environ Resour 33(1):167–185. https://doi.org/10.1146/annurev.environ.33.021407.093240
Simone A (2001) On the worlding of African cities. Afr Stud Rev 44(2):15–41
Singh R (2018) Impact of male out-migration on women left behind: a study of two villages in Uttar Pradesh. Remit Rev 3:75–92. https://doi.org/10.33182/rr.v3i1.427
Sinha H, Kumar A, Xavier Institute of Social Service (Rānchī, India) (2013) Governance of commons and livelihood security. International conference on governance of commons and livelihood security, Ranchi, Jharkhand, India. Xavier Institute of Social Service
Smit W (2021) Urbanization in the global south. Oxford Research Encyclopedia of Glob Public Health. https://doi.org/10.1093/acrefore/9780190632366.013.251
Smith N (1982) Gentrification and uneven development. Econ Geogr 58(2):139–155. https://doi.org/10.2307/143793
Smith N (2002) New globalism, new urbanism: gentrification as global urban strategy. Antipode 34(3):427–450. https://doi.org/10.1111/1467-8330.00249
Smith N (2008) Uneven development: nature, capital, and the production of space, 3rd edn. University of Georgia Press
Steel G, van Noorloos F, Klaufus C (2017) The urban land debate in the global south: new avenues for research. Geoforum 83:133–141. https://doi.org/10.1016/j.geoforum.2017.03.006
Stern BB (1992) Historical and personal nostalgia in advertising text: the fin de siècle effect. J Advert 21(4):11–22. https://doi.org/10.1080/00913367.1992.10673382
Swyngedouw E, Heynen NC (2003) Urban political ecology, justice and the politics of scale. Antipode 35(5):898–918. https://doi.org/10.1111/j.1467-8330.2003.00364.x
Swyngedouw E, Ward C (2020) Producing assets: The social strife of land. Environment and Planning A: Economy and Space 52(2):227–241
Tassadiq F (2022) Producing dispossessed and humanitarian subjects: land acquisition and compensation policies in Lahore, Pakistan. PoLAR polit leg anthropol rev. https://doi.org/10.1111/plar.12506
Tomba L (2017) Gentrifying China's urbanization? Why culture and capital Aren't enough: interventions. Int J Urban Reg Res 41(3):508–517. https://doi.org/10.1111/1468-2427.12494
Turner BS (1987) A Note on Nostalgia. Theory Cult Soc 4(1):147–156. https://doi.org/10.1177/026327687004001008
Ugwu C (2017) History of ethnography: straitening the records. Int J Sociol Anthropol 9:77–81. https://doi.org/10.5897/IJSA2016.0670

Velayutham S, Wise A (2005) Moral economies of a translocal village: obligation and shame among south Indian transnational migrants. Global Netw 5(1):27–47. https://doi.org/10.1111/j.1471-0374.2005.00106.x

Vij S, Narain V (2016) Land, water & power: the demise of common property resources in periurban Gurgaon, India. Land Use Policy 50:59–66. https://doi.org/10.1016/j.landusepol.2015.08.030

Vijayabaskar M (2010) Saving agricultural labour from agriculture: SEZs and politics of silence in Tamil Nadu. Econ Polit Wkly 45(6):36–43

Waley P (2016) Speaking gentrification in the languages of the global east. Urban Stud 53(3):615–625. https://doi.org/10.1177/0042098015615726

Watts M (2017) Political Ecology. In: A companion to economic geography. John Wiley & Sons, Ltd, pp 257–274. https://doi.org/10.1002/9781405166430.ch16

Wildschut T, Sedikides C (2023) Benefits of nostalgia in vulnerable populations. Eur Rev Soc Psychol 34(1):44–91. https://doi.org/10.1080/10463283.2022.2036005

Woolcock M, Narayan D (2000) Social capital: implications for development theory, research, and policy. World Bank Res Obs 15(2):225–249. https://doi.org/10.1093/wbro/15.2.225

Wu J (2008) Land use changes: Economic, social, and environmental impacts. Choices: The Magazine of Food, Farm, and Resource Issues 23(4):1–5.

Yang J, Hui E, Lang W, Li X (2018) Land ownership, rent-seeking, and rural gentrification: reconstructing villages for sustainable urbanization in China. Sustainability 10(6):1997. https://doi.org/10.3390/su10061997

Zaman K, Baloch AA (2011) Urbanization of arable land in Lahore City in Pakistan: a case-study. Can Soc Sci 7(4):Article 4. https://doi.org/10.3968/j.css.1923669720110704.041

Zhan Y (2021) Suspension 2.0: segregated development, financial speculation, and waiting among resettled peasants in urban China. Pac Aff 94(2):347–369. https://doi.org/10.5509/2021942347

Zoomers A (2010) Globalisation and the foreignisation of space: seven processes driving the current global land grab. J Peasant Stud 37(2):429–447. https://doi.org/10.1080/03066151003595325

Zoomers A (2011) Introduction: rushing for land: equitable and sustainable development in Africa, Asia and Latin America. Development 54(1):12–20. https://doi.org/10.1057/dev.2010.97

Chapter 4
Walking the Peripheries: Ethnography as Self and Method

Contents

Abstract This chapter reflects on the methodological journey of conducting multi-sited ethnographic research in the peri-urban east of Lahore, Pakistan, between 2014 and 2022. Approaching the field as a contested and changing landscape, the study traces how urban expansion reconfigures not only land and ecology but also the social and emotional lives of those who inhabit these spaces. Ethnography here is not treated as a neutral technique, but as a journey shaped by positionality, relationships and ruptures. It is not just the making of genuine connections and learning

H. Javaid, *Land Commodification, Urbanisation and Social Change in Peri-Urban Villages, Lahore*, Advances in Asian Human-Environmental Research, https://doi.org/10.1007/978-3-032-10462-5_4

from knowledge holders but also the path of becoming. Fieldwork unfolded across multiple sites and scales, through oral histories, living alongside communities and attending to the rhythms of their everyday practices for the intimate documentation of their day-to-day lives. These methods enabled the capturing of both visible transformations—land acquisition, ecological degradation and displacement—and invisible ruptures, pillage of intangible assets beyond compensation, loss of trust, dislocation, the erosion of community ties and the loss of safety net. By centring on the lived experiences of peri-urban communities, this chapter argues that methodology is inseparable from politics: to write about peri-urban Lahore is to engage with uneven development, dispossession, the fragile ecologies of survival and the fading echoes of belonging. Ethnography, then, becomes a practice of witnessing, listening and tracing life amid loss that is written into the landscape itself.

4.1 Ethnographic Research: Journey to Enlightenment

4.1.1 Background: Trapped in a Shell

Before I map out in detail the methodology being adopted to support the central thesis. I would like to briefly explain how it began in the first place. During MPhil, *Shuhada* town, a small portion of DHA phase 9, scattered on 500 acres (202 ha), 3.5 km off the *Bedian* road, where road construction had just commenced in 2013, was taken as a case study. At the time, thesis was looking at the land-use change and the impact of such housing schemes on the environment with the integration of a small chapter on social impact. During the time, I came across locales and communities which became the motivation for further empirical research for the doctoral book. Little did I know that it not only laid the foundation for doctorate degree but also introduced me to the methodology, I was barely known of, which later became integral for data collection. It was not only for the research but also turned out to be self-exploratory journey simultaneously. The next section explains the choice being made.

4.1.2 Why Ethnography?

It is a multidisciplinary research project that looks at urbanisation and related processes happening on the outskirts of Lahore city through the lens of political ecology but framed under Frank's theory as mentioned explicitly in Chap. 3. Political ecology itself as a critical research field has known the inter-relation and interconnectedness between biophysical and cultural/social world and how they undergo reorganization under different political, economic policies and power structures (Little 2007). Without incorporating the influence of politics, the 'ecology' of a certain environment and its science cannot be understood. As Forsyth (2003) says that science and politics cannot be dealt separately. Therefore, this interdisciplinary field opens a forum that invites researchers from a vast array such as social sciences (anthropology, sociology, geography), natural sciences (e.g. environmental sciences or ecology) or humanities to perceive complex socio-environmental realities by drawing its linkage with power dynamics, inequality and political economy (e.g. globalisation, neoliberalism, economic development) from different aspects (Bryant 1992; Roberts 2020). Similarly, ethnography helps connecting dots by not only enabling a researcher as receptor of information through all five senses but also catching and decoding information emotionally, mentally, physically and sympathetically.

In simpler terms, ethnography is a qualitative research method to gather fine-grained details about people, community in the given space and time and their relation with their lived environment (Jones and Smith 2017). Ethnography as a methodology can have several variants (Hammersley and Atkinson 2010), yet participant observation is central to ethnography (Laurier et al. 2017) where researcher spends considerable time to read, study and observe through sincere interaction with the social groups. Geographers used to practice this methodology in the nineteenth century, but this was not the primary consideration for human geographers for most part of the twentieth century; it was in the early period of the 2000s when ethnography and ethnographic fieldwork became increasingly adopted by political geographers, but by this time, human geographers have also started putting emphasis on ‘everyday power relations’ (Ghoddousi and Page 2020). It is a combination of techniques depending upon situational requirement such as audio recordings, field notes, interviews, etc.

The study is based on multisided ethnography which was first coined by George E. Marcus in 1995 (Marcus 1995). It goes beyond the bounds of single-site ethnographic study due to which the ‘mobile positioning’ allows not only to follow people, material, cultural reproduction and connections across one given space as a vessel but also to follow the flow in more than one sites (as they are significantly continuous yet spatially discontiguous) (Falzon 2016). It has a power to bring forward grounded narratives across multiple geographies. In my research, the multi-sited ethnographic approach was apt as the research site, comprised of several villages, hamlets and urbanised locales, was under continual transition, and people and communities are in the process of social and material landscape reconfiguration.

The following section shows the on-field training for carrying out an ethnography ‘before’ doctoral degree formally began in 2019.

4.1.3 Ethnographer in the Making: Time to Sprout

Purwaningrum (2017) explores how gender shapes one’s positionality and influences the creation of space for knowledge production. My experience as a female researcher in the field brought both unique advantages and specific limitations. Lahore being my hometown was a big advantage for me. It was convenient for me to go to the field, spend a good amount of time and return home by sunset. Nevertheless, in the formative years (especially from 2016 to 2019), I had to face sundry challenges; most of them were because of my shy demeanour and the time I needed to open up which hampered engaging people in conversations. The fieldwork in initial years of the research would not have been possible without my mentors Rabia Nadir and Dr. Munir Ghazanfar, who not only used to accompany me but also taught me immensely right on the field: How to engage in conversation, how to make interlocutors comfortable and how to steer the in-depth interview and whatnot. I also had a language barrier. Language functions as a key instrument for building rapport, allowing people to make acquaintances more easily than someone unfamiliar with the language that enables us to engage with the 'other' by becoming a 'literary insider' (Htong Kham 2024). Punjabi was by far the language of most people; I could understand but could not speak fluently which was a big problem. I realised its importance when people showed reservations in responding to my questions asked in *Urdu* (the lingua franca of the educated urban people) until my mentor repeated the same in *Punjabi*, which at that time reduced my confidence even further. However, those challenges were blessings in disguise confronted in those early days. I spent a couple of years in the field at different intervals and tried to overcome these hindrances that were hampering my journey of research. I started working on language by closely listening to my mentor while she was engaging with people, practicing it at home, with colleagues, at workplace, value-laden reading *Punjabi* scriptures and most importantly through transcriptions that I did in the recorded language and then read them over and over again. That helped me in not only picking up *Punjabi* vocabulary but also the other skills of 'how to decode data' and pin down themes, which was another skill that I had to learn. The other advantage from re-reading transcriptions and re-listening recorded interviews was to develop ability to read between the lines and to pinpoint missing links and aspects that needed to have been given more attention and be further investigated. Other issues were related to the 'ethics' of ethnography which could sometimes make things unseemly and difficult. For instance, seeking permission to record interview or to do it secretly so that the interviewee would not feel hesitant or pressurised always appeared as a conundrum. With time, smart phones became common in villages, and people being very observant tend to get to know that they were being recorded which at times put me in spot. Also, taking pictures covertly, which sometimes used to alert people, could be unfavourable. I, thus, learnt through practice that village people had their own decorum, and tradition must not be breached. Although I had so much to learn, one thing I already had was a sensitive ear and my propensity towards understating the social issues. The rural people in general were very cooperative and forthcoming, but they need to build trust and understand who you are. They do not trust certain sections of the society like the state, the media and the property dealers. You need to assure them you do not represent them, and your project is purely academic.

The next section explains organisation of fieldtrips during degree program.

4.2 Scheduling Fieldworks: Quest for Data Collection

Without fieldwork experience, scholar's feet were never planted firmly on ground. Mukerjee, R.K., 1929, Fields and Farmers of Oudh

Fieldwork is the crucial and indispensable part for fact finding, connects researcher to the real world and provides learning from first-hand experience (Phillips 2017). Theoretical framework and fieldwork are not isolated but complementary to each other. If theory can give direction to fieldwork, without the latter, theory cannot be solidified (Joshi 1981). Hence, to restart data collection, fieldtrips to home country were carefully discussed and planned. As per research plan, three visits in three consecutive years were scheduled to allow me to stay updated with on-going development happening on the fringes which in turn would provide a timeline. Also, spending four to five months on collecting empirical data and then analysing, discussing and sharing it with others helped to identify and see through the blind spots and in sharpening questions for further work. The planning phase was smooth on paper, but execution was not without unprecedented glitches. Supposedly, the first fieldwork, after the commencement of doctoral degree, had to take place from March to June 2020, second in 2021 from March till July and the third and the final field trip again from March till the end of July 2022.

The pandemic of 2021 put a lot of plans into disarray. Back in 2020, when I was leaving for fieldtrip, the Covid was hardly any threat in Pakistan, but one week post landing in Lahore, circumstances changed. It was not just the lockdown but the fear of contracting this deadly disease that halted the research work completely for which I had travelled all the way from Germany. Since I could not go to field and due to the uncertainty of the situation, instead of extending my trip, I decided to fly back. Going back means that I had to wait till the next year hoping things would get better. However, the whole first year of PhD went down the drain as I went back empty-handed. In 2021, my plans for field trip again became a huge matter of concern, and I was told to reorganise my research plans to exclude fieldwork, so that my health and safety would not be compromised. Nonetheless, it was practically not an option, and I wanted to take a chance by all means. Hence, after much discussion, I managed to get the approval, and I flew back to Pakistan with the intention of being in field no matter what. Fortunately, the way empirical data collection panned out, it was worth the risk. Nonetheless, how to draw a starting line became a labyrinth; the succeeding section sheds lights on it.

4.2.1 From where to Begin to Data Outpour

Discussion session with my former faculty fellows at the Lahore School of Economics (LSE) was the point of departure. In the light of our discussion, I decided to start my work from Lahore School where many drivers come from the same area.

LSE helped me in finding people from different hinterland as most of the staff used to come from the nearby villages. Especially Mr. *Tariq Gill* and Mr. *Basharat* went out of the way to broaden my data collection through fieldwork. They took pains to help me unconditionally because they themselves became interested in the nature of research only on the basis of the acquaintances we developed over the long haul. On the other hand, I learnt along the way that my village respondents themselves wanted to be heard and seen, regardless of gender and age. There, I realised that ethnographic research offers way more than I had imagined. Apparently doing nothing for the ones from whom I was collecting knowledge and oral histories but offering an ear and to let them know and feel that they were heard, noticed and valued turned out to be priceless for them as interlocutors mentioned this at several encounters. People in those villages, trapped in a process of development, were immensely willing to share their side of story of transition that how things had been changing in both physical and social realm.Longitudinal ethnographic research with multiple field trips was thus conducted from 2014 to 2022; however, in the following section, I will be specifically highlighting the fieldwork from 2020 onwards, where interlocutors were not subjects but real knowledge producers regardless of age, sect, ethnicity and gender, whose time and information is considered as gift (Falzon 2016).

1st fieldwork	March to July 2020	Pandemic
2nd fieldwork	March to July 2021	5 months
3rd and final fieldwork	March to July 2022	5 months

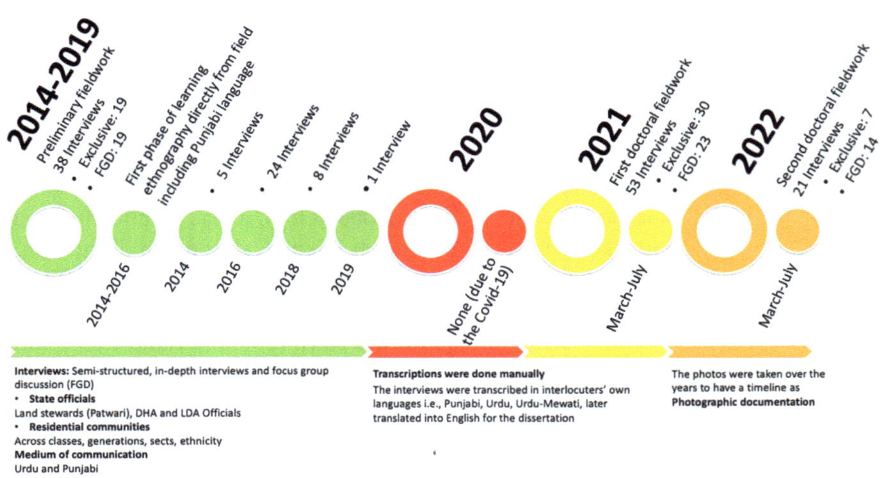

The younger ones limned the present, and the middle generation not only depicted the comparison of their younger selves with today's generation but also helped me realise the reasons for intergenerational difference of opinions and how maturity, and past experiences opened a window to see the real assets of life. The aged men and women were real encyclopaedias of untold, unrecorded tales of history, migration, partition turmoil and change over time, not just in landscape but in relations, warmth, trust, network, community spirit, food, water and so on. They looked like temples of cognition and wisdom. They were the keepers inherently of intangibles—a value system not attached with money, to make them last. They were also the analysers of the past, today and tomorrow and critics of their own. Therefore, ethnography was used as a primary method for this collection of empirical data from 40 peri-urban villages on southeast of Lahore, located at and around two main connecting roads *Barki* and *Bedian*. Little (2007) says, 'political ecology as an ethnography'. This deliberate method of choice is the foundation of the research to unpack the social complexities tied with spatial-temporal changes and to catch the nuances. Ethnography as a methodology contributes in understanding urbanisation that can help policy makers, planners and practitioners (Koster 2020) because these processes, meaning and experiences are not homogenous or universal but vary across place and space (Herbert 2000). My positionality as a female ethnographer privileged me to document 'change' through rural/peri-urban women's perspective which is the integral part of the research. The following section will break down ethnography into sub-methods which were used to investigate the research question.

4.3 Fieldwork

The eminent Indian social scientist P. C. Joshi in his ethnographic experience sums it up beautifully:

> After my fieldwork I, too, was not the person I was before. I was a haunted being — haunted by the people among whom I had lived and of whose joys and suffering I had had direct experience. When I think of my villages, I am both hopeful and fearful of change, because change has now ceased to be an impersonal process for me. It has assumed a personal meaning as a source of joy and suffering, hope and despair, for persons whom I had known intimately. (Joshi 1981: 475)

4.3.1 Field Interviews

The in-depth interviews are a subset of ethnography because this very method persuades a researcher to join people in their natural settings to become a part of their lived environment (Barwick and Miller 2023). As decoloniality of knowledge also through ethnography research strongly suggests to have an alternative way of understanding 'other' according to 'how communities see and do', it is one way of respecting interlocutors' experiences and knowledge (Manning 2018). It teaches a researcher to be utterly respectful towards other people's time but more of their pace and wavelength and allow space to build trust and confidence. The relationship between land and people has drastically changed with the creation of the market in reshaping and restructuring the peri-urban environment. Ethnography is the method of choice given its holistic approach to delve and fathom the multifaceted realities of socio-spatial change of people' natural environment. This research methodology offers a creative, reflexive and interpretive process through continuous and iterative fieldworks (Tanchareon 2005). It is the full immersion and submersion of researcher for developing relation with people and places who are not just subjects but the knowledge producers. Being a participant who constantly observes, participates in and processes information from people's surroundings is essential to unfold impacts of urbanisation on communities. 'How' people have been experiencing change is a fundamental question that can debunk various myths circling around like 'cash is tantamount to peace and happiness', 'development has improved lives' and that 'nostalgia' of the past is overrated and makes people delusional. This research has shown how research methods from different fields may be used for transdisciplinary research projects without tagging them as methods of a particular discipline.

The snowball technique helped conduct 112 interviews,[1] including telephonic interviews—38 interviews (2014–2019) as preliminary work; 53 interviews (March–July 2021), 20 interviews (March–July 2022) and 1 distant interview in 2024. The in-depth, semi-structured interviews were conducted from 2014 to 2022 at 40 plus locales.[2] Fifty-six focus group discussions and 56 one-on-one interviews were conducted at multiple sites primarily in Punjabi and Urdu languages, while a few confabs were the amalgamation of *Mewati/Urdu* or *Mewati/Punjabi* dialects, with denizens across generations, class, ethnicity, religion and gender, intertwined with participatory observations. During the process, I learnt how to evolve the preparation for fieldwork to have more productive results. I often try to find the connection first and also tend to get familiar with the interlocutor beforehand through the informant. I started spending time on writing questions that could help gathering more data or relevant themes in terms of similarity, disagreements, contradictions or other surprises. But once I reached the people' lived environment, I let them take me through their lives and environment. Yet, keeping mental record to which aspects I needed to redirect the conversation is important. At the places which were all new to me, I tried build rapport first to get the ball rolling. These one-on-one engagements were recorded in mobile phone, in order to preserve each and every word as it was told and spoken. It is of paramount importance to save dialogues so that one can listen to them repeatedly as their words in local language provide a unique depth and detail which a casual listening could miss.

4.3.2 Transcribing Interviews

Afterwards, these recorded interviews were converted into written text which was a toilsome and time-consuming process. For the matter, I did not use any software for audio conversion or data analysis. I did the job manually. Although there are

[1] List is attached to the appendices as Appendix B.

[2] List is attached to the appendices as Appendix A.

software for the job, many researchers still opt for traditional way of manual transcriptions that provide more precision and accuracy to data (Tang 2023). I also took help from one of my acquaintances to share some load. For that, I am deeply grateful to Mr. Nadeem who generously helped me transcribe audios to text. As there were several long interviews, therefore, in order to save time, the recordings which could not be transferred on pages completely, I heard and re-heard those recordings and transcribed the excerpts of the interviews which fit into themes. Although this exercise took plenty of time, the advantage of doing manual transcriptions was that it helped immensely for deep immersion into data to get meaning and more insight to pay heed to details which otherwise could be overlooked (Bird 2005). The transcriptions (written version of interviews) played an essential role for data analysis by extracting themes and not only to underline contradiction if there is any, but also to allow researcher to re-visit interviews with aim of reading between lines for better understanding and interpretations (a blend of what was told and what was seen and observed). Since I have not used any computerised software for data analysis and done it myself, after making transcriptions, which were handwritten in a separate journal. After audio to text conversion, I spent time in reading interviews to see which information I could get and what were the areas that needed to be probed. Those extracted themes were taken as 'manual codes', for instance, water, food, inflation, development, money, land, military, coercion and so forth. It was important for me to take into account 'who' was talking (class, gender, age) and their natural setting (village, urbanised locale, ghetto, walled-off locale). This helped me to keep my biases aside and to fathom the 'truth' from the interlocutor's perspective. This is where art of listening plays a vital role (in collecting data) that an ethnographer can offer, which is way more than participant observation (Gerard Forsey 2010). I agree to this fact as I experienced the same while I was in Christian village called Durgpura during a 2-hour-long interview with a group of men in Dera adjacent to Haveli (a cattle-keeping place). Dwellers showed trust on me by being brutally honest about the process taking place, the role of government, the army, marginalisation and vulnerability due to the social group they belong to and how the class difference was accentuated by the role of money through commodification of land and the story of loss (mostly intangibles along with health, food and access to clean water) which was entwined with their individual material gains. When I asked their names and mentioned, '*Your identities will be kept confidential*', they unanimously replied, '*We don't care Madam, even if you write our names, all we want is to be heard, we have no clue what are you capable of in terms of bringing betterment but we can't thank you enough for coming here and for lending an ear, it is more than enough for us*'. Valediction was fraught with prayers for me, but a need to be recognised, seen and heard was undeniable, and there I realised what services an ethnographer can provide unknowingly to the community even if they are ephemeral with lasting impact. I encountered the same in several different occasions. The transcriptions or recorded versions proved to be trips down the memory lane which did help not only in refreshing my memory but also in developing and strengthening a connection with people I spoke with. I also taught myself by going back to interviews about posing questions and how to indulge people in a conversation to find answers without putting words to their mouths and also allow them to talk me through their lives and stories of change, the way they liked.

For the interviews, I tried to sample as many categories of people as possible. Among officials and professionals, I interviewed state officials such as people working at revenue record keeping department, police officers, lawyers, (indirect conversations with) architects, activist, LDA officials working at city planning department and involved in making Lahore's masterplan 2050 and DHA officials. As part of the methodology, I also conducted interviews with local businesspersons, including property dealers. Within this category, two types of realtors emerged: those who were formerly large landlords and transitioned into the land business long ago, and those with smaller landholdings who adopted this profession more recently. In addition to real estate actors, individuals involved in milk-based businesses were interviewed to understand the role of agrarian-linked enterprises in the local economy. The main category of interviewees, however, comprised residential communities. These included long-standing groups of neighbours who have lived in the area for generations, often linked by kinship, as well as more recent residents who moved from different villages or cities. Among these diverse groups, the Mewati or Meo community stands out as a distinct group due to its different ethnicity and language; they migrated together as a clan from across the border during Partition and are historically regarded as an 'outsider' community within the region. Community can further be divided because of religion such as Christians (minority) and Muslims (majority); Muslims are then divided into two main sects—Shia and Sunni Muslims. However, gated community is another form of community which is supposedly an upgraded version of living, which was taken into consideration especially in terms of those who used to live in a village and then moved to relatively upscale locales. The group of people who live farthest to the east of Lahore, I name them as hybrid communities. Further classification was done on the basis of socio-economic background or, in simpler words, 'class' and gender as men and women have different encounters with development-led change across three sets of generation (teenagers, middle-age group and old).

4.3.3 Field Notes

Field notes are means to record observational data on field, which are usually handwritten. However, it is not a straightforward activity and also not possible to capture everything while on field (Hammersley and Atkinson 2010). The habit of taking field notes was another important part in the process which I started developing in the later years of the research. Jotting down what one had seen and observed on the field while being there in person helps in keeping record of observations without losing them. However, keeping journal and a pen while talking to people without any prior contact and reference sometimes became a hurdle as much as recording interviews without bringing to their notice in the beginning. Through my personal experience, I realised that people became stonewalled or highly hesitant while talking which hampered the process of building trust. It is of no surprise that why villagers especially in the area who had been facing development get more sceptical at first; at times, they misunderstood me as DHA personnel, occasionally as a journalist; however, this very thought that came to their mind at times worked in my favour because they wanted to have their story being heard. With women from walled-off villages particularly, it was often difficult to record them either as notes or tapes, if I went up to them without being accompanied by any known face from their locality.

One of such anecdotes was when I went to Chachuwali (a walled-off locality, where arable land had been taken already). I went there with no prior contact and hence spent days in forging social relations. After a few days, I went back and saw a different atmosphere which was less welcoming, hesitant and reluctant. I even felt being avoided by the ones I was familiar with. I was perplexed and constantly questioning where I went wrong, what exactly must have happened in the past few days and how to fix it. Due to my gender, I had the advantage that I could talk to women freely which was one integral source of data collection through in-depth conversations or interviews. Finally, my anxiety came to an end when one of the former interviewees who was an elderly woman let me in her house again and had a talk briefly. While walking me to one of her neighbours, she eventually expressed, *'People are little suspicious about you because you come here, sit with us, record us in your device and leave'*. Though it made me a little nervous at first, I was happy at the same time because now I knew how to mend it. I spent a considerable time with women, re-introduced myself and my purpose of visiting the village, explained a need for recording interviews and opened the floor for them to ask me what they wanted to for their satisfaction which was important to win their trust back. During that time, I kept my notepad, a pen and a recording device all in bag. After a thorough inquiry, they hugged me, welcomed me again and treated me as one of their own that made me rest easier. I had to rely on mental notes though, and soon after getting out of the place, transferring what was stored in memory onto a notebook was a priority. Women already living in enclosed areas by the DHA tend to look for an answer in response to, '*What can you do for us?*'; however, women on the fringes were more open and amicable.

The above vignette is taken from my personal experience. The purpose of sharing it as an example is to call attention to a couple of important aspects of fieldwork. This not only underscores the challenges a researcher can encounter while doing ethnographic research but also zeroes in on the disregarded fact that having research site in one's own country does not make a human geographer an 'insider'. Since I had been going to villages for a long time which made me more conversant with people, their surroundings, dos and don'ts, how to dress myself up in order to blend in, a language that is pivotal in creating a zone where people wouldn't hold themselves back from wearing their hearts on their sleeves and most importantly giving them time and space to trust and to get myself aligned with their pace out of utter respect, I often tried to find a contact beforehand, but there were times when I randomly entered a home without knowing anyone and started making acquaintances which also put me in odd predicaments. The villagers with strong community ties can easily identify an 'outsider' from a distance. I was native to the city but an 'outsider' at the same time to rural areas. As Mughal (2015) also established through his writing that being home-grown doesn't save one from being labelled as outlander. By dint of my gender, while corresponding with office personnel, people working in court, police officers, revenue record holders, etc., I found it easier to talk to them in order to access contacts or data. In fact, there were instances where security guards did not let me stand in a male queue and escorted me directly to the relevant person and waited there till I was done. In rural communities, men in villages were very respectful, and out of regard, they stopped by just by noticing that I might be looking for someone and then gave me their time generously and contacts to continue doing my work. In subsequent visits, they would accompany me to introduce me to other relevant contacts in the village. On the other hand, women in general went through gamut of emotions and stream of thoughts from being cautious and highly inquisitive about nature of my work and what I was doing on streets in scorching heat, which reflected in their guarded and stern vocal tone. Gradually, this evolved into a communicative style marked by refinement and courtesy, coupled with appreciative expressions, after seeing someone came to their door to know 'how are they doing?', 'how do they "feel" and so on. I also noticed that young women who were in their in their twenties were thrilled and enthusiastic while having me at their place because they found me different from them in regards to social standing and educational background, as few mentioned out loud.

Nevertheless, it was not all rosy. Safety and security are the foremost concerns when it comes to females roaming in places they hardly know. It happened in my case as well. I was always escorted by a trust-worthy person the whole time, and I used to have my own conveyance to reach places where I had to conduct fieldwork. However, I avoided making my way into highly urbanised, crowded settlements by walking. There were instances where I felt scared and harassed even if I was alone for hardly five minutes, as those few minutes seemed like years. Due to research purposes, I had to exchange mobile numbers for further inquiries and further references. I was in uncomfortable situation at times when I had to deal with such entities who tried to breach work ethics. For this reason and also because of easily commuted access, I decided to spend day on field but not nights. This may have given me a setback to an extent, but this is why I sometimes thought that if only I were a male, things would have been different; yet, I can't deny the far greater perks of being a female researcher.

It is important to understand and acknowledge the fact that knowing the same language and developing connections with local communities do not make a researcher as one of them. It may cultivate characteristics of both insider (due to familiarity) and outsider (which the researcher is), which is termed as 'in-out-sider' (Htong Kham 2024). In fact, a researcher 'researching at home' or away stays in migratory state the whole time with added pressure of recognition, where local/traditional knowledge from the home country is perceived as 'researcher being delusional' if it does not align with imperial western knowledge. Mignolo (2000) calls for 'border thinking' as a decolonial practice that fosters the coexistence, negotiation, and recognition of multiple knowledge systems and epistemologies. Rather than suppressing or marginalising alternative ways of knowing through ongoing colonial logics, border thinking insists on epistemic plurality and challenges the dominance of Western modernist frameworks. Escobar's (2018) work further deepens this argument by proposing that decoloniality must be grounded in place-based ontologies and relational modes of life that modern development continually undermines. His idea of "designs for the pluriverse" highlights the need to create worlds in which diverse forms of knowledge, including indigenous, agrarian, and local ecological epistemologies, are valued as co-equal rather than subordinated to technocratic or capitalist models of progress. Together, Mignolo and Escobar show how decolonial thinking opens pathways for reimagining environmental care, sustainability, and community life in ways that honour the worlds and knowledges that dominant development paradigms have long obscured or erased. Mignolo (2000) calls for 'border thinking' to encourage, navigate and coexistence of multiple knowledge systems, epistemologies instead of suppressing and marginalising the 'other' way of thinking by continuous colonisation.

4.3.4 Time to Reflect - Writing and Returning to the Field

Writing field notes is only the first step in ethnographic research. Equally crucial is the return to these notes at the end of each day, sitting alone with the day's movements, interactions, and observations recorded in the field journal. This quiet act of revisiting the day's moments is more than procedural. It is a small pause where the noise of the field settles, and impressions begin to take shape. Such reflective anchoring is essential to systematic data logging and sustained immersion in qualitative inquiry (Emerson et al., 2011; Wolcott, 1994). It aligns with the understanding of the researcher as an instrument who must remain self-aware, responsive, and critically attentive throughout fieldwork. Sitting with the material each evening, and placing it within its broader social and ecological context, allowed the initial chaos of encounters to become legible. The constant flow of information that arrives through multiple senses can easily become overwhelming (Spradley, 1979). Through iterative reading, memo writing, and slow interpretation, patterns began to surface, silences grew noticeable, and gaps in the narrative opened possible directions for the next set of conversations (Saldaña, 2021; Charmaz, 2006; Mohajan & Mohajan, 2022; Black et al., 2021). These sessions ensured that fieldwork was not a hurried accumulation of data but an ongoing dialogue between experience and understanding. Recent methodological discussions have reshaped how ethnography is imagined today. Continuous residence in the field is no longer seen as the only path to depth. Instead, ethnography often unfolds through cycles of immersion and withdrawal, where periods of close engagement are followed by intentional time away to process stories, restore energy, and reflect ethically on one's role. This rhythm reflects an emerging recognition in the discipline that depth does not require unbroken proximity, and that distance can sharpen attentiveness. Ideas such as "committed localism" speak to this shift. Ethnographic depth can be achieved through focused and ethically attuned engagements that honour the field while acknowledging the researcher's practical and emotional limits (Skovgaard-Smith, 2024). Stepping back becomes part of the work of thinking, feeling, and staying accountable. Rather than diminishing immersion, these pauses help sustain it by preventing exhaustion and enabling clearer, more compassionate seeing. The growing scholarship on researcher wellbeing reinforces this perspective. Temporary withdrawal from the field is not detachment but a necessary form of ethnographic care, allowing researchers to return with greater clarity, humility, and steadiness. In this framing, caring for oneself is inseparable from caring for the communities one studies. Reflection, rest, and recalibration become integral to ethical engagement. Together, these rhythms of presence and distance deepen the quality of ethnographic work. Immersion offers access to lived textures and meanings, while withdrawal creates space to understand how those moments fit together. Moving between the two ensures that the ethnographer remains attentive not only to what unfolds in the field, but also to how they interpret and represent the lives entrusted to them. The integrity of the work rests in this oscillation, where both proximity and pause matter in equal measure.

4.3.5 Discussion a Method

Discussion as Method While structured interviews, field notes, and systematic observation formed the core of the methodological process, the discussions that followed became just as crucial for making sense of the evolving field. These conversations unfolded at different moments and across multiple mediums: spontaneous exchanges after field visits, occasional telephone calls, and later long Zoom sessions with discussants in Lahore after my return to Germany. Each of these dialogues opened interpretive pathways, pushed me to revisit assumptions, and drew my attention to dimensions I had previously overlooked. In many ways, these discussions acted as intellectual hinges, moments that unlocked my thinking and allowed the research to move forward with greater clarity. These exchanges resonate with methodological traditions that view analysis as dialogic, collaborative, and iterative rather than a solitary act (Gergen, 2009). The back and forth conversation itself became a form of dialogic analysis: meaning was not merely extracted from data but co-constructed through ongoing dialogue that illuminated the multiple ways participants interpreted and made sense of their changing world. Rather than treating the field as a set of static observations, interpretation unfolded as an active and relational process. At the same time, many of these conversations functioned as peer debriefing (Lincoln and Guba, 1985; Creswell and Miller, 2000). Trusted interlocutors, whether discussants in Lahore or colleagues in Germany, challenged emerging interpretations, questioned assumptions, and prevented premature analytic closure. These debriefing moments helped identify blind spots, strengthened reflexivity, and added credibility to the findings. They acted as a safeguard against over interpretation and ensured that analysis remained ethically and empirically grounded. These discussions also embodied forms of collaborative interpretation, where participants themselves shaped the analytic direction. Their insights, disagreements, anecdotes, and concerns guided the questions I returned to, the themes I expanded, and the theoretical frames I reconsidered. Instead of analysing about communities, the process evolved into thinking with them, allowing their interpretations and lived logics to inform the categories and arguments built in this study. Taken together, these dialogic, collaborative, and reflective engagements show that discussion was not a supplementary activity but a methodological axis around which interpretation revolved. They provided the emotional and conceptual space needed to process dense stories, sift through overlapping meanings, and hold together the complexity of lived experience. Through these exchanges, the analytic process became a shared endeavour, strengthened by many voices, shaped through relation, and grounded in the collective labour of thinking alongside others.

4.4 My Positionality

It is imperative to talk about the positionality especially from the gender and class perspective. Undoubtedly, I have an advantage that Lahore is my birthplace. I had been going to villages for a long time which made me more conversant with people, their surroundings, dos and don'ts, how to dress myself up in order to blend in, a language that is pivotal in creating a zone where people wouldn't hold themselves back from wearing their hearts on their sleeves and most importantly giving them time and space to trust and to get myself aligned with their pace out of utter respect. I often tried to find a contact beforehand but there were times when I randomly entered a home without knowing anyone and started making acquaintances which also put me in odd predicaments. The villagers with strong community ties can easily identify an 'outsider' from a distance. I was native to the city but an 'outsider' at the same time to rural areas. As Mughal (2015) also established through his writing that being home-grown doesn't save one from being labelled as outlander. Being a female researcher, my experience revealed that gender gave me an advantage. Office personnel, people working in court, police officers, revenue record holders etc., I found it easier to talk to them in order to access contacts or data. In fact, there were instances where security guards did not let me stand in a male queue and escorted me directly to the relevant person and waited there till, I was done. Best interviews were done in villages where I was considered trust-worthy and people felt safe while talking to me. Men in villages were very respectful and out of regard, they stopped by just by noticing that I might be looking for someone and then gave me their ample time generously and contacts to continue doing my work. In subsequent visits they would accompany me to introduce me to other relevant contacts in the village. I also believe that I could talk to women comfortably and spend time with them because my gender gave them comfort, although I found them more inquisitive about nature of my work and what I was doing on streets in scorching heat but also very appreciative of the fact that someone came to their door to know 'how are they doing?'. I also noticed that young women who were in their in their twenties were thrilled and enthusiastic while having me at their place because they found me different from them in regards to social standing and educational background, as few mentioned out loud.

Nevertheless, it was not all rosy. Safety and security are the foremost concerns when it comes to females, roaming in places they hardly know. It happened in my case as well. I was always escorted by a trust-worthy person the whole time, I used to have my own conveyance to reach places where I had to conduct fieldwork. However, I avoided making my way into highly urbanised, crowded settlements by walking. There were instances where I felt scared and harassed even if I was alone for hardly five minutes, those few minutes seemed like years. Due to research purposes, I had to exchange mobile numbers for further inquiries and further references. I was uncomfortable situation at times when I had to deal with such entities who tried to breach work ethics. For this reason, and also because commuting offered a safer and more practical option, I made the decision to spend my days in the field but not my nights. I recognise that this choice created limits in how deeply I could immerse myself in village life. Night-time rhythms, spontaneous gatherings, and many informal interactions remained partially out of reach. Yet these constraints were not simply logistical. They were profoundly gendered. There were moments when I found myself thinking that if I had been a man, the field would have unfolded differently. My presence would have been less scrutinised, my movements less restricted, and my access to certain social spaces far more open. A male researcher could have stayed overnight, moved freely, and blended into the village landscape in ways unavailable to me. Rooms, conversations, and after-dark worlds would likely have been accessible without hesitation or negotiation. Rather than interpreting this as a methodological deficiency, I came to understand it as an essential analytic insight. Feminist scholars remind us that fieldwork is never neutral and that research relationships are shaped by the researcher's embodiment and the social order within which they move (England, 1994; Rose, 1997; Abu-Lughod, 1990; Behar, 1996). These factors influence not only access but also what is spoken, what remains unsaid, what is shown, and what is withheld. In this sense, my inability to stay overnight illuminated the gendered structures that organise rural social life and the guardedness that surrounds women's mobility after dark. This awareness did not weaken the research; it strengthened it. It revealed the power dynamics that shape not only the daily lives of the communities I studied but also the very conditions under which knowledge is produced. Feminist methodology emphasises that reflexivity is integral to analysis rather than peripheral to it (Haraway, 2013; Harding, 1991). By acknowledging the limits created by gendered norms of safety and propriety, I was also acknowledging the wider cultural landscape in which my research was embedded. The separation between day and night in my fieldwork became a window into the separation between who can move freely and who cannot, whose presence is unquestioned and whose requires explanation, who is protected and who is policed. These patterns are not merely methodological obstacles; they are data. They signal something fundamental about gendered access, vulnerability, and the moral geographies that structure everyday life in rural Pakistan. Seen through this lens, the methodological challenge of not staying overnight becomes part of the story of the field. It is a reminder that ethnography is not simply about being there. It is also about recognising why one cannot always be there, and what these limits reveal about the world being studied. My partial access becomes a form of insight. My distance becomes a way of seeing. My gendered constraints become an analytic resource rather than a barrier.

4.5 Limitations

I faced immense difficulty in getting access to data that was specifically related to the DHA, and most of the times, I ended up failing at getting hands on this. Also, while I was on field in the area where landscape was changing, settlements were walled off and especially if the area was disputed because settlers did not want to vacate land, there I found people extremely reluctant to talk, mostly those who occupied the lower ranks in hierarchal order. Security guards who worked there also tend to get suspicious, and at one instance, guard did not hesitate to scare the woman off whom I was talking to. The access to revenue records was also a tedious job as Mauzas were divided amongst different people, and those people were often not so generous when it came to sharing details. Sometimes, they were ready to talk but did not want me to record or take photographs of the records. Similarly, with the archival records also, although the archives have been well organised or at least responsible members have been doing their best to keep the records secure, yet as for me, the records from the 1990s were unavailable due to peculiar reasons, and getting access to them was a long process to which relevant personnel suggested you to look for an alternative. Therefore, relying totally on fieldwork for collecting as much information as I could was challenging and also instilled fear in me if I was doing enough to do the justice to all those people who gave me even five minutes of their time. Like, I mentioned in the above section that I did not go to places where I could not feel comfortable and secure; hence, I would say that gender did limit me at certain occurrences. The Covid-19 did affect a schedule, and in order to compensate and to keep up with the timeline, I had to manoeuvre research plan.

4.6 Remote Sensing[3]

Multi-temporal satellite imagery was used to visualise the urban expansion and agricultural land loss between 1965 and 2023. The usage of Corona data, due to its high spatial resolution and which dates back to the 1960s, in combination with

[3] I again would like to thank Dr. Susanne Schmidt for her time and efforts in developing these maps.

Landsat-data and Sentinel-2 data, has proven to be particularly suitable for the investigation of urban expansion (Dame et al. 2019; Hussain et al. 2023). Therefore, a high spatial resolution panchromatic Corona image from 1965-10-4 (spatial resolution 0.6–1.2 m), a multi-spectral Landsat 5 scene from 1988-02-23 and a Landsat 7 scene from 2000-03-19 (both with a spatial resolution of 30 m) and a Sentinel-2 scene from 2023-02-21 (spatial resolution of 10 m) were downloaded from earthexplorer.usgs.gov.

Furthermore, to visualise the expansion of the built-up area since the nineteenth century, the world settlement footprint (WSF-DLR) was used for the evolution from 1985 to 2015 and 2019 (geoservice.dlr.de). Furthermore, the urbanised area before the 1980s, classified into pre-colonial, colonial and postcolonial, was digitised based on the map drawn by Schimmel and Mayer (1982). The location and outlines of the DHA phases were derived from DHA plans which are available online (dhalahore.org). All maps were georeferenced, and the outlines were manually digitised on screen. Village names and locations were derived from Open Street Map, Google Earth and the historical topographical map (sheet NJH 43-2, series U502) in the scale of 1:250,000 drawn by the US Army Map Service and compiled in 1955.

4.7 Secondary Data

Most of the secondary data was taken from the literature available in form of research articles, newspaper archives, books, DHA official websites and social media such as Twitter and YouTube. The newspaper editorials were also the integral source. *Iqbal Qaiser*, a Punjabi historian, has been the first-hand source who munificently provided me data from his lifelong work on Punjabi villages. The other source was poems, folklores or poetries on rural life which were collected from different sources such as directly from poets or those available on social networking sites on rural living.

Hence, this chapter not only intended to accentuate the methodology being used to conduct research but also tried to break down elements of certain methodology to explain how empirical data was gathered.

References

Abu-Lughod L (1990) Can there be a feminist ethnography? Women & Performance: A Journal of Feminist Theory. 5(1):7–27.

Barwick D, Miller B (2023) Lessons from the field: an ethnographic approach to understand people's lived experience. User Research in Government. https://userresearch.blog.gov.uk/2023/11/30/lessons-from-the-field-an-ethnographic-approach-to-understand-peoples-lived-experience/

Behar R. (1996) The vulnerable observer: Anthropology that breaks your heart. Beacon Press.

Bird CM (2005) How I stopped dreading and learned to love transcription. Qual Inq 11(2):226–248. https://doi.org/10.1177/1077800404273413

Black GB, van Os S, Machen S, Fulop NJ (2021) Ethnographic research as an evolving method for supporting healthcare improvement skills: A scoping review. BMC Medical Research Methodology. 21(1):274. https://doi.org/10.1186/s12874-021-01466-9

Bryant RL (1992) Political ecology: an emerging research agenda in third-world studies. Polit Geogr 11(1):12–36. https://doi.org/10.1016/0962-6298(92)90017-N

Charmaz K (2006) Constructing grounded theory: A practical guide through qualitative analysis. Sage.

Creswell JW, & Miller DL (2000) Determining Validity in Qualitative Inquiry. Theory Into Practice, 39(3), 124–130. https://doi.org/10.1207/s15430421tip3903_2

Dame J, Schmidt S, Müller J, Nüsser M (2019) Urbanisation and socio-ecological challenges in high mountain towns: insights from Leh (Ladakh), India. Landsc Urban Plan 189:189–199. https://doi.org/10.1016/j.landurbplan.2019.04.017

Emerson RM, Fretz RI, & Shaw LL (2011) Writing Ethnographic Fieldnotes, Second Edition. University of Chicago Press.

England K (1994) Getting personal: Reflexivity, positionality, and feminist research. The Professional Geographer. 46(1):80–89.

Escobar A (2018) Designs for the pluriverse: Radical interdependence, autonomy, and the making of worlds. Duke University Press.

Falzon M-A (2016) Multi-sited ethnography: theory, praxis and locality in contemporary research, 1st edn. Routledge. https://doi.org/10.4324/9781315596389

Forsyth T (2003) Critical political ecology: the politics of environmental science. Routledge

Gerard Forsey M (2010) Ethnography as participant listening. Ethnography 11(4):558–572. https://doi.org/10.1177/1466138110372587

Gergen KJ (2009) Relational Being: Beyond Self and Community. Oxford University Press.

Ghoddousi P, Page S (2020) Using ethnography and assemblage theory in political geography. Geogr Compass 14(10):e12533. https://doi.org/10.1111/gec3.12533

Hammersley M, Atkinson P (2010) Ethnography: principles in practice (3. ed., reprinted). Routledge

Haraway D (2013) Situated knowledges: The science question in feminism and the privilege of partial perspective 1. In Women, science, and technology (pp. 455–472). Routledge.

Harding S (1991) Whose science? Whose knowledge? Thinking from women's lives. Cornell University Press.

Herbert S (2000) For ethnography. Prog Hum Geogr 24(4):550–568. https://doi.org/10.1191/030913200100189102

Htong Kham S (2024) Outsider from within, insider from without: negotiating researcher positionality in comparative social research. Int J Qual Methods 23:16094069241254004. https://doi.org/10.1177/16094069241254004

Hussain A, Schmidt S, Nüsser M (2023) Dynamics of mountain urbanisation: evidence from the Trans-Himalayan town of Kargil, Ladakh, India. Land 12:920. https://doi.org/10.3390/land12040920

Jones J, Smith J (2017) Ethnography: challenges and opportunities. Evid Based Nurs 20(4):98–100. https://doi.org/10.1136/eb-2017-102786

Joshi PC (1981) Fieldwork experience: relived and reconsidered. The Agrarian society of Uttar Pradesh. J Peasant Stud 8(4):455–484. https://doi.org/10.1080/03066158108438147

Koster M (2020) An ethnographic perspective on urban planning in Brazil: temporality, diversity and critical urban theory. Int J Urban Reg Res 44(2):185–199. https://doi.org/10.1111/1468-2427.12765

Laurier E, Miller R, Muñoz Zech DI (2017) Ethnography (pp. 9780199874002–0171) [Dataset]. https://doi.org/10.1093/obo/9780199874002-0171

Lincoln YS, & Guba EG (1985) Naturalistic Inquiry. SAGE.

Little P (2007) Political ecology as ethnography: a theoretical and methodological guide. Horizontes Antropológicos 3

Manning J (2018) Becoming a decolonial feminist ethnographer: addressing the complexities of positionality and representation. Manag Learn 49:135050761774527. https://doi.org/10.1177/1350507617745275

Marcus GE (1995) Ethnography in/of the world system: the emergence of multi-sited ethnography. Annu Rev Anthropol 24:95–117. https://doi.org/10.1146/annurev.an.24.100195.000523

Mignolo W (2000) Local histories/global designs. Coloniality, subaltern knowledges and border thinking. In: Local histories/global designs: Coloniality, subaltern knowledges, and border thinking. https://doi.org/10.23943/princeton/9780691156095.001.0001

Mohajan D, Mohajan HK (2022) Constructivist Grounded Theory: A New Research Approach in Social Science. Research and Advances in Education. 1(4):8–16.

Mughal M (2015) Being and becoming native: a methodological enquiry into doing anthropology at home. Anthrop Notebooks 21:121–132

Phillips R (2017) Fieldwork (pp. 9780199874002–0158) [Dataset]. https://doi.org/10.1093/obo/9780199874002-0158

Purwaningrum F (2017) Reflections on fieldwork: a comparative study of positionality in ethnographic research across Asia. https://doi.org/10.31219/osf.io/df3n4

Roberts J (2020) Political ecology. The Open Encyclopedia of Anthropology https://www.anthroencyclopedia.com/entry/political-ecology

Rose G (1997) Situating knowledges: Positionality, reflexivities and other tactics. Progress in Human Geography. 21(3):305–320.

Saldaña J (2021) The coding manual for qualitative researchers.

Schimmel A, Mayer J (1982) Lahore. Entwicklung und räumliche Ordnung seines zentralen Geschäftsbereiches. Die Welt Des Islams 22(1/4):159. https://doi.org/10.2307/1569818

Skovgaard-Smith I (2024) Ethnography beyond the tribe: From immersion to "committed localism" in the study of relational work. Journal of Organizational Ethnography, 13(2), 233–249. https://doi.org/10.1108/JOE-05-2023-0025

Spradley JP (1979) Interviewing an informant. The ethnographic interview. 46:55–68.

Tanchareon S (2005) Ethnographically informed community and cultural assessment research systems (EICCARS) working paper series. Basic Classical Ethnographic Research Methods Secondary Data Analysis, Fieldwork, Observation/Participant Observation, and Informal and Semi-structured Interviewing. https://www.academia.edu/32722719/ETHNOGRAPHICALLY_INFORMED_COMMUNITY_AND_CULTURAL_ASSESSMENT_RESEARCH_SYSTEMS_EICCARS_WORKING_PAPER_SERIES_Basic_Classical_Ethnographic_Research_Methods_Secondary_Data_Analysis_Fieldwork_Observation_Participant_Observation_and_Informal_and_Semi_structured_Interviewing

Tang R (2023) Manual transcription. In: Okoko JM, Tunison S, Walker KD (eds) Varieties of qualitative research methods: selected contextual perspectives. Springer International Publishing, pp 295–302. https://doi.org/10.1007/978-3-031-04394-9_47

Wolcott HF (1994) Transforming qualitative data: Description, analysis, and interpretation. Sage.

Chapter 5
Invisible Ruptures, Visible Ruins: Life, Land and Loss in Peri-Urban Lahore

Contents

Abstract This chapter explores the profound socio-ecological transformations taking place in the peri-urban landscapes east of Lahore, Pakistan, between 2014 and 2022. Drawing on historical accounts, oral histories and ethnographic fieldwork, it examines how state-backed real-estate expansion—particularly the Defence Housing Authority (DHA) projects—has reconfigured agrarian ecologies, everyday livelihoods and community relations. What was once a fertile, canal-irrigated agrarian landscape embedded in reciprocity and collective ties has been progressively commodified, fragmented and absorbed into an urban order centred on speculative value and elite aesthetics. By analysing both the visible changes (land acquisition, soil mining, displacement, loss of agriculture) and the invisible transformations

H. Javaid, *Land Commodification, Urbanisation and Social Change in Peri-Urban Villages, Lahore*, Advances in Asian Human-Environmental Research, https://doi.org/10.1007/978-3-032-10462-5_5

(erosion of trust, cultural dislocation, psychological insecurity and loss of intangible community assets), the chapter highlights how neoliberal urbanism extends beyond material landscapes into cognitive, emotional and social realms. The findings foreground peri-urban villages as more than sites of 'developmental transition'; they are contested spaces where agrarian modes of life are undone, identities reshuffled and ecologies destabilised. Ultimately, the chapter argues that urban expansion in Lahore is not simply spatial growth but a political project of dispossession, one that externalise ecological costs and undermine localised forms of resilience. It calls for recognising peri-urban spaces as critical frontiers of environmental justice, where struggles over land, ecology, and community futures unfold in uneven and unequal ways.

5.1 Introduction

The research aims to study socio-ecological change in the peri-urban landscape of East of Lahore city through a historical account of recent developments. This study started in 2014 and extends till 2022, the last year of field research, and this chapter presents the findings from 2014 to 2022. Since 2000, this area particularly grabbed the attention for housing area investment and infrastructure building which opened up the market in real estate. The land lust has caused immense mutation in physical and social realm. This chapter divides results in two broad categories, i.e. physical and social change eventually transforming environment. Nevertheless, this very division was to see the change in tangibles which were accessible to naked eye and how this impacted the real assets in form of intangibles which were the products of a certain ecology and shaping the communities the way they were prior to the advent of development.

The southeastern periphery once dotted with several (peri-urban) villages composed of large swathes of fertile agricultural lands irrigated by BRB canal and its distributaries. One of the differences between border villages (farthest to the east) and peripheral villages is that border villages only use tube wells for irrigation and do not have canal water, whereas most of the southeast area is canal irrigated. Since canal water is much better for the crops, it provides strength to the soil and better yield (Lambardar-Pathankey, 2021). The agrarian land envelops settlement area which is called *Laal Lakeer/Lal Dora or Red Tape.* During the British Raj, a red line was drawn on maps to demarcate village population separate from the farmland; it was also used by the revenue department to divide *Abadi* from productive land.

The part of the village, i.e. inhabited area (comes under *laal lakeer*), does not have a land entitlement document or registration unlike for privately owned lands. Hence, this particular part has a collective ownership. Therefore, when people want to sell the area under red tape, it is usually done by a mutual decision, and amount is divided equally amongst them. The peri-urban villages are connected with two main historical roads *Bedian* and *Barki*—formerly known as *Harikay* Road (SO[1]-Durgpura, 2021). Many of these villages were before partition such as *Jahman, Durgpura, Natha Singh, Heera Singh, Hudiara, Sarja Marja, Dhira, Padhana, Mandianwala* and *Bhangali*. Other settlements, such as Chachuwali, were born after partition. It was inhabited by the migrant *Mewatis* who after crossing the border stayed at Walton and later moved to *Liddarh,* but due to their distinctive language and ethnicity, they got into clashes with local clans; eventually, they came to *Chachuwali* (Chachuwali, 2021; DHA Broker-Chachuwali, 2021). The villages comprise different communities such as *Mewati* or *Meo*, *Rajpoot, Malik, Awaan, Jatt, Arayi, Gujjar* and *Rahmani;* artisanal castes such as *Saqqay, Taili, Naee* and *Julaha*; weavers; *Mochi* (Cobblers); *Lohar* (Ironsmith); *Tarkhan* (Carpenters); *Muslim Sheikh* (lower caste Hindu who converted to Islam) and *Easaee* (Christians), and *Bhangali* has *Shia* and *Sunni* communities too.

Figure 5.1 is the virtual representation of socio-ecological change in pixels. The urban sprawl is rapidly moving towards the east and has been taking over agricultural lands from poor localised communities and jeopardising them in the name of development. This form of development, driven by a state-captured real estate regime in which DHA functions as the primary and most influential actor, operates with uncontested authority and an economic reach that renders it effectively unanswerable to any oversight. In this configuration, DHA becomes both the executioner of the city's agrarian landscapes and the chief architect of aestheticised enclaves designed for elite consumption. The agricultural land was not just an immortal piece of earth that has potential to feed and sustain, but agrarian mode of production was a way of life where cash and market economy were far behind, but reciprocity, pay-in-kind and interdependency were the foundation. Most of the left half of the map is the built-up area, and on the right side, farthest to east, close to the border is still standing as a rural realm. The map shows the expansion of DHA phases and its sectors in leap-frog fashion, consuming several historical villages. The villages as mentioned earlier are at different stages of the development, but it has marked the ones that no longer exist to provide a record about villages that have been lost in the process and the locales that have not been displaced, but urbanism has overwhelmed the area. The commodification of land and the resulting urban sprawl continue relentlessly under neoliberal policies that champion such infrastructural development projects. This transformation is not confined to the conversion of agricultural land into real estate. It signals the onset of a far broader set of environmental, social, and ecological disruptions, all of which are completely externalised in the compensatory packages offered to cash-poor farmers. Already positioned as marginal and disposable, these farmers absorb multilayered losses that no monetary payment can meaningfully address. The map shows the land-use change and pins the localities, which can help in understanding that places do not change in isolation, and material change leads to change on all three landscapes—cognitive, psychological,

[1] The codes are used to identify interlocutors. The detailed list of interviewees and their coding is attached to Appendix B at the end of the thesis.

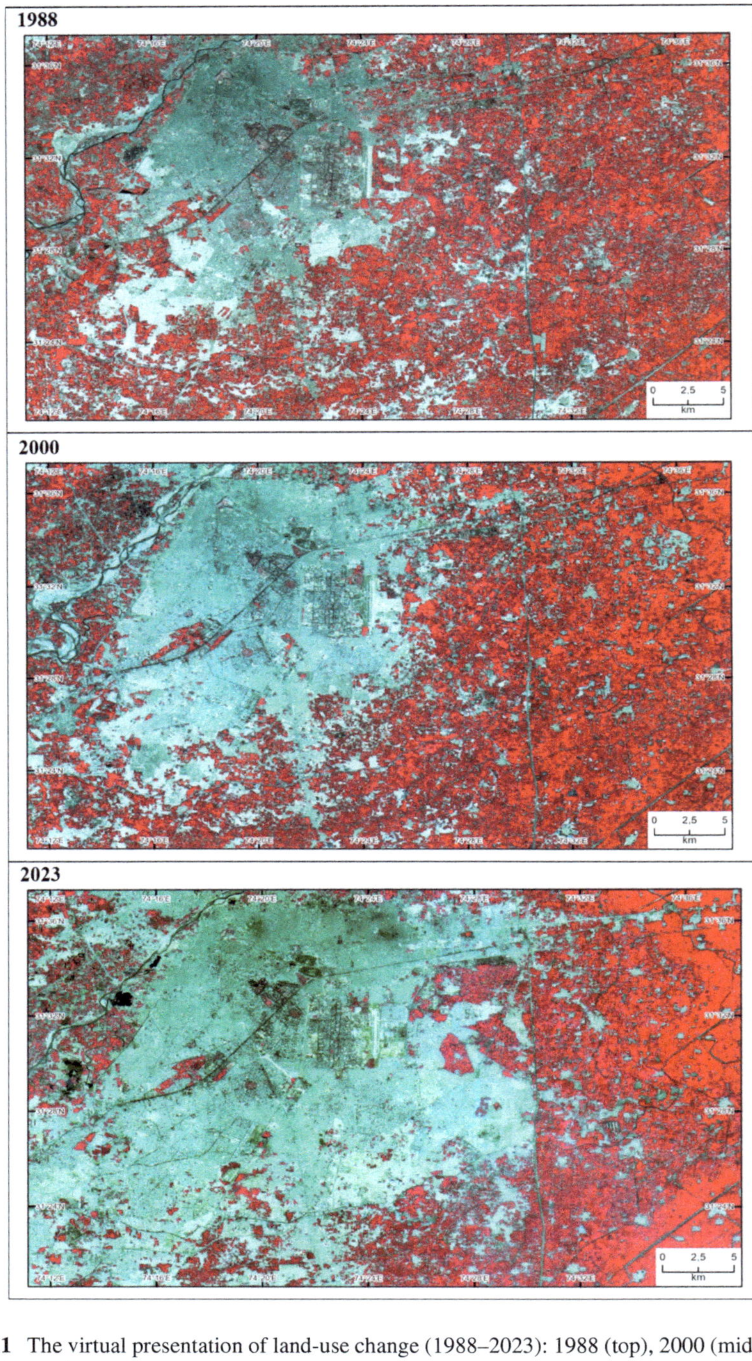

Fig. 5.1 The virtual presentation of land-use change (1988–2023): 1988 (top), 2000 (middle) and 2023 (bottom)

emotional. The villages and settlements being shown in the map are not just the physical spaces that are going through reconfiguration and reformation but also has been experiencing class shuffling, in situ migrations, change in community due to uninvited encroachment of so-called development that is class centric and callous to environmental and natural resources. A development has converted every resource with use value to exchange value including the intangibles be it trust, reverence, connections, community, network, power of togetherness, security, fear, coping tools in the time of need and whatnot.

The region used to produce cash crops and several vegetables such as wheat, rice, corn, fodder, peas, carrots, tobacco, potatoes, tomatoes, radish, leafy green vegetables and fruits such as guava, oranges, sweet lime, etc. As mostly people were involved into agriculture, what they produced was also being sold in Lahore city through nearby markets or *mandis* such as in *Koth Lakhpat, Bhagwan Pura* and *Badaami Bagh*. Farming and livestock were almost tied together; therefore, besides vegetables, the area had provided milk also. Hence, alongside farming, the milk business served as another important source of livelihood for local residents. In terms of landholdings, most villages contained a mix of small landholders and landless households, while large landlords were either absent or represented by only a few families. Those without land had long relied on tenant cropping as their primary occupation, a livelihood that sustained them until DHA's acquisitions began in Charrar in 1977 and initiated the sale and conversion of agricultural land, then from *Jalal abad* in 1996, followed by *Sangatpura* and the rest (see Appendix A), land acquisition picked up pace in early 2000s, some villages sold their land in around 2006 such as *Chachuwali, Durgpura, Gohawa; Natha Singh* village was sold in 2011, *Saduwala's* land was taken in 2012, *Pathanwala's* land was purchased by DHA in 2013 and the buying is still going on to date. DHA buys parcels and gathers the land first, and then, it starts transforming it; therefore, a timeline for land acquisition can be observed. This is why, the study site was able to show a whole picture of change as these villages were found at different stages of the development, yet occurring simultaneously.

After the partition, villages were divided between Lahore and Amritsar districts, and each region got 85 villages which were there before independence with the centuries' old history, e.g. *Padhana, Pathanke, Lehna Singh, Mota Singh, Natha Singh, Liddharh* and so on (FG-DHA Land providers, 2024). Further, it was clarified that since then '85 villages' which was at that time, a number of pre-partition villages but now used as the identification of the area as the villages in numbers have surpassed 85. During the current data collection, these villages were at different phases of take-over development.[2] From the relatively pristine border villages through the walled-off villages and villages reduced to a slum or ghetto provided the journey of a village on the path of development, leading eventually to the complete removal of villages such as *Sangatpura, Gobindpura, Jalal abad, Haveliyan* and *Heera Singh*. The next section answers, 'What is a village'?

[2] See an Appendix D.

5.2 A 'Village'

The following excerpt has been taken from the multiple meetings and long conversation with a *Charrar* resident, who was 50 years old when we met for the first time. He was born and raised there and saw the place changing its shape and colour in front of his eyes. By profession, he was a barber, who now had shifted his shop in Defence. This village will remain of importance as it was the first village from where acquisition started through one-on-one dealing with DHA, which also set the precedent for the DHA to never come to the forefront.

'Charrar was 80% of shamlaat and 20% of the ownership. We did not have big landholdings, so we were tenant croppers. What used to happen was that people who were doing cultivation as tenant/share cropper, they did not have their own land, after partition once Sikhs left, land became under the ownership of those who were already cultivating it. There were large tracts of agricultural lands and people like us who lived around farmland, our bread and butter were associated with that land. Although we did not have a big land but we were into agriculture, now the same land is buried under DHA, where there is K-Block water tank, we used to have our 'Dera' there, I still remember myself playing there. I remember it very clearly. DHA started acquiring land from this village in 1977 and by 1980 DHA started building. This very area where DHA has been built over, was once extremely fertile land. Everything we used to produce from fields and there was availability of everything, you name it and you have it. The wise elderly people resisted it, they went to General Zia's office and told his staff to find a barren land for construction. This particular area used to produce pea that entire Lahore ate. Now 250 g of pea costs 100 rupees (0.33 €), which was once available almost gratis. Where there is now DHA phase 5, people used to grow high quality carrots and one could smell it from a distance, it was considered the house of carrots, now you can no longer find that kind of vegetable. Even landless never slept empty stomachs, food was never an issue even if somebody did not own the land. You see, when you grow pea, you can grow another two to four vegetables at the same time, radish pods on the boundary wall of the field, these were of certain height that also provided protection as nothing could enter the field, along with it, we grew gourds. With pea, okra and coriander were also produced, and in the center, we grew carrots and saag (green leafy vegetable) in winters. Whether people owned land or not, they could go outside and pluck the desired vegetable for cooking. Then we had big houses and now we are cramped into marlas (25.3 m^2). Before DHA, we had big houses with haveli (a separate area to keep the animals). When I got married, all guests at the wedding were accommodated at home, and were served food at the same place, now if someone dies, we cannot accommodate ten people because house sizes have shrunk'.

Village fairs hugely contributed to the recognition and identity of the place, and the generations who spent time in a village setting still use farmer's calendar and festivals to specify dates and events. He continued, *'A village fair happens every year even till now from 2 days or one month of Bhadun*[3] *(15th August-14th*

[3] See Appendix C to have a look at Punjabi months.

September), we still have the same zeal and excitement for the village fair and this excitement is way more than our Eid festival. When this event arrives, all our married sisters, daughters, come back to their homes without any invitation (their arrival requires no invitation), drummers especially participate, festival has its own color and vibrance. One can learn a lot from the event, as some people present drama, sing Sufi songs, other one would recite Waris Shah's famous poetry Heer, attendees could gather and learn historical things. Then, there are stalls selling local produce. It brings so much joy and happiness to us, it is almost like a son's wedding' (Charrar, 2016; 2021).

The border villages were of interest to have a picture of what a village looked like. Villages like *Jahman, Sarja Marja, Padhana* and *Pathanke* are amongst the places where a semblance of the above described still looks closer to reality of rural culture. This helps a great deal in imagining peri-urban villages and their agrarian ecology prior to disruption. They are located farther of the east, close to the Indian border. People have not left agrarian mode of production and are still tied to milk and land for crop production. Apart from seasonal *Rabi* and *Kharif* crops, dwellers produce vegetables according to their will and the need for personal consumption (see Fig. 5.2).

Fields are watered through tube wells, and hand pumps are also used to acquire drinking water from the ground. In order to protect the quality of groundwater, the wastewater ponds (see Fig. 5.3) are built and cared after by the communities.

Livestock is the other source of livelihood and nutrition (see Fig. 5.4). Peasants keep a portion of the milk they produce for domestic use and sell the rest in the market. However, for those who are into selling milk business, the trend has shifted to selling milk in entirety for money.

The village and its land are multipurpose that accommodates all (see Fig. 5.5).

Fig. 5.2 Fertile agricultural lands of rice (left) and wheat crops (right)

Fig. 5.3 Wastewater pond, Pathanke, 2021

Other than owner, farmers' cultivators are either share croppers or contract farmers who pay upfront rent in cash to the landlord. Electricity is available, but the need for fuel is met through woods and mostly from cow dung cakes (see Fig. 5.6), although gas cylinders are now also coming into use.

The rural setting offers security to women and young girls which is completely amiss in urban setting and urbanised localities. Some women are also involved in money-based economy by doing jobs in the city. But women who are living in villages are not left behind when it comes to earning. They are indulged in doing craft work at home. Some are into making ornaments from different materials and selling them, while others do embroidery and stitching which are also receding in villages now. Hence, houses are also cottage industries in which women actively take part. Other businesses by women include the opening of small shops next to their house. I have also seen woman as an owner of the vegetable shop who runs it with her husband. The shop was located in Charrar, on a crowded street, yet she felt comfortable as she was surrounded by people who were either the community or relatives. They take the orders for the embroidery work from a middleman who buys it from them. They also make other household items like strand and weed hand-held fans, plates and containers to keep bread (see Fig. 5.7).

Women also make earthen hearths and *'tandoors'* to cook and make bread at home wherever clay or earth is used as construction material be it wood-burning hearth (see Fig. 5.8, on the left), plastering, kitchen shelves (see Fig. 5.8, on the right) or making a raised terrace for seating and work. The craft and design are quite apart from any needed agricultural work in the fields and care of animals and a variety of household chores.

Fig. 5.4 Communal 'Dera' for keeping animals, Pathanke, 2021 (top) and open communal grazing land, Jahman, 2019 (bottom)

People have big houses that accommodate culture of joint family. Such villages are still practicing reciprocity and the culture of sharing with strong community ties where dependence on people has still not been replaced by cash.

Another layer can be seen through villages which are on military farmlands (a glimpse of a border village in Fig. 5.9); they are more 'hybrid' cultural settlements that have several similarities with above-mentioned villages, but they are also participating in modern-world money-making opportunities predominantly through rental incomes and non-farm jobs. They are closer to main roads which gives them accessibility and connection to the city. Residents of such settlements still get grains and dairy from farms and cash coming from house tenants and family members employed in city jobs. Examples are *Burj*, *Chughalpura (Habibabad), Jindri*, *Jindra Kalan* and *Rakh Padri*; they are changing materially but trying to keep the intangible cultural intact.

Fig. 5.5 Space shared by all: Boys are taking a bath in canal (left) which is also used by cattle to take a bath (right)

Fig. 5.6 Mud houses, cow-dung cakes for drying, Sarja Marja, 2021 (left); drying and preservation of dung cakes for heating and cooking (right)

Rural world is not a utopia. It is also not a perfect world free of worries and discomforts, yet people call it a heaven, *'City can never be a place even close to what this village is'* (FW/2-Sarja Marja, 2021). At another instant, a middle-aged man standing in front of his house with his relatives as neighbours, amongst those who did not sell the land, whereas most of the village had been bought by DHA, had been negotiating and eventually locked a deal, said, *'We are asking for 16.374 € / 2.5 km^2, if they offer half of it why would we leave our heaven'* (Chak Dhira, 2022). Nonetheless, the priceless intangible assets are the product of history and ecology. Once the ecology is gone, it takes away the hidden treasures too. Therefore, the next section is an attempt to elucidate, 'what is the ecology of a village in this area'?

Fig. 5.7 Women's handicrafts: (hand fan in the making, embroidered pillow cover, breadbin (left) and fancy embroidered cloth piece (right (2021); trinkets in the making by a woman in Kohriyan, 2016 (bottom)

Fig. 5.8 Kitchen in rural houses with a gas stove made by women (left) and kitchen shelves made from mud (right)

Fig. 5.9 Sunset in a border village—Sundown (somewhere) in rural ecology, 2021

5.2.1 Ecology: Smells Like 'Home'

Away from the hustle and bustle of the city, driving far to the east, a road covered by shadowy trees was taking me to manifest the 'rural'. The main road led to the unpaved bumpy path surrounded by greenfield swinging with wind. I was engrossed in witnessing a different realm that was otherwise only heard in conversations—quiet, serene, green, open but covered with vegetation, gold wheat and fodder sprouting by tearing chest of mother earth, under the blue sky filled and mixed with yellow coming from fiery sun. The perfect blend of colours all in the nature's palette. Finally, a car stopped, and voices brought me back to look meticulously at the rural culture still prevalent in village *Sarja Marja*, the last village of Lahore, which shares border with two Indian villages *Chhina* and *Narli*. *Jatt, Malik* and *Arayi* are three castes found in this village. *Pathanke* is *Meo* village as only *Mewatis* reside there. The account touched upon below was recorded during a field visit in 2021 and 2022 to the mentioned border village due to which other border villages such as *Padhana* and *Pathanke* and the ones on military farmlands such as *Burj, Chughalpura* and *Jindra Kalan* were also validated.

I felt being invited by the place to see myself the sustained ecology and 'almost-textbook' rural world. Despite my reserved demeanour, I soon found myself as one of them, a feeling that is amiss in urbanised, modern and especially elite areas of the city. *'We live in a paradise, here we have everything, we grow vegetables which other people can also pluck from the ground, above all, we feel free to go anywhere within a village, there is no fear in walking alone as I experience in the city'*, a young girl broke the silence though her words as if she heard my thoughts while my gaze was wandering all over the place (FW-Sarja Marja, 2021). *'Let me take you to the tube-well, it is closed for now but it is our favorite picnic spot in summers. My sisters, friends and I take stroll in the fields in our free time which is very invigorating for us, it is our safe haven, therefore our male figures don't have to accompany us'*. She took pride in iterating the security which rural setting offered them, especially when I stopped after seeing a group of young boys coming from the opposite side on a narrow dirt road, they moved past us without raising their eyes which was uncanny as compared to harassment females experienced in so-called developed areas of the city. She took me to her home where her parents were already standing at the door to welcome me with their open arms. I felt humbled not only to get to know them but also to see that humans and space they share create an environment. It reminds me of Cecelia Ahern's quote 'Home is not a place, it is a feeling', a researcher who was an 'outsider' yet had a sense of belonging. None of the gates had locks, and women were inside of the houses not only as guardians but to keep their homes open for others. None of the members asked my names at the entrance unless they took me inside with great respect, and I introduced myself without being asked. *'Curd, yogurt, milk, ghee, butter etc., they are all home-made, we never get short of milk as we have animals'*, one of the women pointed out to the pot placed on low fire to make yogurt (see Fig. 5.10).

Fig. 5.10 High variety of food in the rural area of Lahore varying from home-made curd (bottom), mustards seeds for oil production (top left) to beans, coriander and mint leaves (top right)

More people joined in once I stepped out of the house to take a tour of the village. An army of young, well-behaved, sensible boys, young girls and two middle-aged men took me on the tour of the village. There were open spaces where aged men with their *'hookah'* sat with others, having a chitchat, and some gathered to spend quality time outside of small shops in the late afternoon. Milkmen were getting ready for 4 pm ritual of milking cows in order to send it to the main city. An unmetalled village roads between open spaces on the one side and mud houses on the other. While on a walk, children were playing old traditional games (unsupervised by their mothers but watched over by several sets of eyes). The shared culture was reflected not only through common-animal much more keeping places but also through their personalities. I hardly heard 'my, mine', but 'we, together' were the much more common pronouns. As I was told by one of my respondents, *'When people live together even big issues are solved whether it is match making, arranging dowry, they stand together in grief or tragedy, the grievance is shared, the happiness is shared'*. People are still tied to agricultural mode of production, living as tenants on military land which is not permitted for financialisation. They keep two parts of the produce, and one goes to military. Their other occupation is selling milk. It is not cash-free culture, but most of the basic needs, especially quality food and a community to ensure safety net, brotherhood and manpower, keep their physical, emotional, mental and financial subsistence and stability intact. People in border villages still produce their own food on a parcel of land where they grow crops (see Fig. 5.10). These are mostly those who are the tenant croppers, *'We have got a land on contract, we pay rent annually which is 30-35K (98-114 €), where there is an availability of irrigation water, they pay 40K (131 €) per year'* (FW-Sarja Marja, 2021).

Apart from the availability of land to grow which gives them power to produce and maintain subsistence, how else they cherish the form-of-land the most is the roof over their head. The normal architecture of the house in a village (see Fig. 5.11), regardless of geographical placement, is to have a spacious veranda surrounded by smaller rooms as people preferred to have ample space to facilitate others (Harpalke, 2018; Charrar, 2021). The houses are large enough to have ample space for relatives and friends to get together in an open area surrounded by rooms. *'We live in heavens'*, words I heard from many different persons, *'look at the fields, feel the breeze, one does not miss fans, we still love to sleep on the roof at night, talking to each-other across walls and rooftops without needing a gadget. Our men might not earn much but still they have kept us with respect and care, we have a lot to cherish. We can't even imagine to migrate to city'*, a woman from (FW*N-Sarja Marja, 2021) added with whom I had a candid conversation on the low wall of a rooftop, a vantage point to savour picturesque view of the village. Similar account was delivered at another military farmland 5 years back, where women appreciated the openness of rural living not just in space but in hearts too (FW-Rakh Padri, 2016). And not just for the people, as they consider animals and livestock no lesser than their family, they make sure that they have ample space for keeping animals also. It is common in villages that the houses are big and in complete harmony to adjust human and non-human family members. For them, they are their children who need same

Fig. 5.11 Traditionally large houses (mud houses in first two rows in Sarja Marja, 2021); House in Durgpura, 2022 with modern touch (bottom left); Animal-keeping space in house in Chughalpura, 2022 (bottom right)

amount of attention, love, affection and the level of commitment and responsibility as a human child needs and deserves. Luckily, I had the opportunity to witness first-hand the relationship women maintain with their livestock. I was sitting in the Mewati village of Chachuwali, speaking with a group of Meo women, when I sensed a sudden tension in the air. The eldest woman in the household seemed restless, her body language shifting as our conversation continued. Curious, I gently asked what she was feeling. She replied, with a mix of worry and anticipation, "My cow is in labour, today is the time." Mewati women and young girls were all waiting to welcome a new member to their household which was a calf as cow was in labour. Since, young females do not take active part in looking after cattle, and especially during birthing time, ladies of certain age (regardless of their marital status) are not even allowed to mark their presence while cow is giving birth. Same goes with men at home. Hence, there was a separate room for her where eldest woman in the house who was constantly checking up on her, caressing her, calming her down and was herself going through a concoction of emotions (thrilled, worried, eager, anxious) called out other elderly women in her neighbourhood through walls, who started getting together. There was a strange silence inside the house where young women (married and unmarried) were waiting for the news to arrive. It took me time to swallow that it was not her daughter but that was how she was treating her cow. Before bidding farewell, I hugged and asked, *'how should someone congratulate at this time?'*, she smiled at me and replied, *'just like when a human-child comes into the world'*. The conviviality of people from villages be it at any stage of transformation is still 'rural' in nature, i.e. filled with warmth and happiness despite the geographical placement, which they carry with pride.

The above-mentioned text was to give an idea about a village regardless of its geographic location. Be it farthest to the east, touching Indian border where defence considerations do not allow mainstream development enter or the peri-urban villages in southeast Lahore around two main roads. The now-urbanised locale or milieu-in-transition were once very similar to border villages. The subsequent sections illustrate how urbanisation in peri-urban villages as a development policy that floods cash into these marginalised but self-sufficient realms has been causing change. It also gives a reality check from ground realities and local people's reaction about whether it is a progress or regression. And what is the purpose of displacing and replacing the current population with an upper class rich low-density population at huge cost.

The next section draws a picture of change that is very evident through material transformation when one is travelling from ecologically sustained areas to places that carry overwhelming symbols of modern development and material change. It is like a time machine that enables a person to visit the past and compare it with contemporary times.

5.3 Making of a New Landscape

'Welcome to market-island', I said it to myself when I drove back from the eastern side of Lahore to the peripheral area, lying on southeast. The landscape shifted abruptly: carpeted roads, newly laid infrastructure, manicured verges, decorative roundabouts, and monumental entrances announced a different world. This transformation begins with DHA's promise of modern and aesthetic living, a vision that reshapes the periphery into a curated display of urban aspiration (see Fig. 5.12)

Fig. 5.12 DHA as a real-estate developer versus environment: DHA's promise for the rich (top left), Invincible—Defence Golf Resort (a new DHA project) (top right), Multifaceted relation between canal and locals (bottom row)

meant only for the ones who can afford it. The development project aiming for the elite by the state institution is the encroachment in the life of a local community. One of the most visible forms of interference is through creating space for general public that is linked with common resource with multiple perks, which breaches the privacy of these rural communities. For instance, *Shabbir Sharif* Road alongside *Khaira* Distributary (a long water channel, shown in Fig. 5.12, where women and young girls still come to wash cloths, and male members of the vicinity use the water channel to wash their bikes and rickshaws) leads straight to *Barki* Road. The distributary is sandwiched between two busy roads and fully exposes women. Women were seeing being conscious with their *dupatta* (a fabric to cover head) while sitting on relatively steel slope for laundering, and some were using the same fabric to cover half of their faces as veil. Men accompanied them to ensure their safety, standing vigilant at a shorter distance, keeping sharp gaze at every passer-by (be it in a vehicle or on foot). Localised communities also carry invisible walls of privacy around them, safeguarded by their culture and traditions. Trespassing by realtors, and the steady infringement on villagers' land in the name of "modern development," is not merely an encroachment but a political incursion. It is a barging-in that violates the sanctity of lived spaces, a forceful reminder of how power operates when capital decides who belongs and who must move.,

The Barki Road projects two starkly contrasting faces of "development" (see Fig. 5.13): one dismal and eroded, the other lucrative and money-generating, where prosperity is narrowly measured through the lens of economic boom. The pictures down below were taken on site at different periods; chosen ones tell the story of

Fig. 5.13 Market in land

landscape from going through change to become a new face. The road in between is used to separate place in two halves; on the left side of the road, there were several villages, privately and state-owned. Now, the most common sight is the placement of banners, billboards and ground-mounted signboards talking about the construction of phases or extensions and their upcoming projects. The fleet of loaders and excavators digging ground, heavy sound of bulldozers to level the terrain need no introduction and dust-blinding sites are the new faces of the area which was once quiet agrarian realm. The productive land is disappearing, with white walls with red strip standing as a symbol of power display and DHA's sovereignty. The other side depicts the business hub. It is a commercial area swamped with a number of real-estate offices, developers' offices, international food and coffee chains such as Gloria Jeans. The picture on the left shows the journey where heart of the earth was crushed to build the imported modern symbols of development.

The merciless scrapping of a 1.5 m thick layer of top soil, cutting down of trees and orchards and clearance of land are preparatory steps for the construction and extension of DHA phases and sectors. An old cattleman lamented the loss of nature's gifts and State's shallow priorities while he was grazing his cattle in the middle of DHA phase, *'Ten, ten trucks used to stand here for loading mangoes and guavas and were delivered to the city, and now there is nothing'* (Bhangali, 2021). Loss of agricultural lands and the butchering of fruit trees for erecting concrete jungle are amongst the forgotten treasures that are never compensated. The military-backed-up business venture, popularly known as DHA, ensures secure, comfortable living and high-end living for the elite in a trailblazing gated housing. However, the land over which these symbols of modernity are shining comes from peri-urban communities who produced and sustained themselves as well as many others through a plethora of tangible and intangible assets. Relatively sustained ecologies have been reduced to degraded service sector, caught up in everyday struggles of life. A well-established property dealer from *Haer* (2016) summed up the process of land transaction, how and for whom it worked, in these words, *'Rich are getting richer and the poor are becoming poorer, because the wealthy take away everything which once belonged to the poor'*. Following collage (see Fig. 5.14) shows the route of development via agrarian economics to create the 'aesthetically beautiful'.

Fig. 5.14 Life cycle of the agricultural land in the epoch of neoliberalism. (**a**) Point of departure. (**b**) Brutal killing of the mother earth. (**c**) Business knows no bounds. (**d**) Work in progress unapologetically. (**e**) Emergence of new face

Urbanisation on the periphery of built DHA has entered the village to create psychological turmoil also. The amount of fear is associated with the class denizens belong to. The photo (see Fig. 5.15, top) is the depiction of one such occurrences, while I was roaming around and saw a woman from a distance with her cow and calf. The ground was uneven and looked like hills that woman and her livestock had to climb and cross after every few steps. She was a Christian woman, holding a wooden stick in one hand and constantly looking over her shoulder. As I wanted to talk to her, I began walking through terrain in an attempt to meet her halfway. I though did get a

Fig. 5.15 Life between 'maze' and 'amaze'—people are striving to 'remain', 2022

Fig. 5.16 Behind the walls—world transforms, Chachuwali, 2021

chance to have a brief conversation during which she told me that she was from Jhallaran, and soon this village would be erased (Jhallaran, 2021). She was talking to me, but her gaze was all over the place. All of a sudden, a man who was a security guard shouted, *'What are you doing here, go away'*. She ran and vanished within seconds. Moments later, he stopped me too and asserted DHA's authority through casual interrogation. It was another reminder of how private power claims public space through fear, policing, and exclusion. He further revealed that where we were standing right now, these were all agricultural lands, but DHA took over, and now, the villages behind the wall (Jindra kalan, Jhallaran) would be erased surely. At another instant, a woman with her children was plucking fodder from the ground, and she got scared when I got closer to her, and without getting her job done, she gathered her children and left immediately. It made me realise again that fear is real and legitimate.

There is a lot more to the story that is not apparent on the surface and needs to be unwrapped, despite what can be observed through naked eye. Tangible change is easy to fathom, but this is not all, it is just a beginning of social change—physically, psychologically, emotionally, morally and traditionally, it is a shift from one world to another without moving an inch, and economically, by restructuring and reshuffling of class structures. Roots of 'change' are deeper than what they look like (see Fig. 5.16), which is used as a metaphor of such intangible chaos.

5.3.1 How Does This Happen?

The urbanisation is predicated on creating an insatiable craving for 'space' and 'place' in 'time' to build new structures such as road networks (see Fig. 5.17), housing schemes and communication networks by dismantling the old landscapes. Research site has undergone massive transformation by transforming agricultural

Fig. 5.17 Mist of progress—'blurred' the sight, 2022

lands to DHA posh housing society, its extended phases, commercial areas, bridges and whatnot. All of this is happening by upending the agricultural lands and displacing and enclosing the previous settlements or the complete erasure of peri-urban villages.

The underlying processes that enable this physical transformation begin with land acquisition. This acquisition takes multiple forms, whether through the seemingly straightforward purchase of land parcels or through more covert methods such as scraping and removing topsoil for use elsewhere. Land is also seized by transforming or reclassifying graveyards, shrines, and other commons including pathways, schools, mosques, wastewater ponds, and grazing lands. These takeovers are facilitated not only through bureaucratic manoeuvring but also through tricks, deceit, intimidation, scuffles, and, at times, outright violence (Villagers, 2014–2022).

5.3.2 *Land Acquisition*

'DHA has raised the value of land,' a realtor claimed (FM-Haer, 2016); *'People are becoming millionaires overnight',* these are some of the most unheard narratives from property dealers. According to them, it is simple: people are selling lands and in return they are receiving money (according to land acquisition law). They also realise and point out that there is no rule or law but the law of the money *'There is no law in here, only one who has money is successful. There is no resistance over selling of land and why would they have resisted. In fact, on the other hand people compare their respective land deals: he got 2 crore,*[4] *I would demand 3. How could DHA face difficulty, people took money and are spending it, not in thousands but in crores'* (Bhangali, 2021). It is also a common belief that people happily and

[4] 1 crore PKR = 32,698 €.

wishfully sell their land, but at the back of their minds, there is a realisation. *'DHA will take over this land in any case and once they map out their phases, whatever appears on the surface, they will remove everything. People have small holdings but DHA is ready to buy more than they can offer. People are happy to sell and DHA is more than happy to purchase'* (Natha Singh, 2022).

DHA does not make all payment for acquired land in money but also a share of housing land. However, after the *Charrar* incident, DHA does not come in the open to lock the deal; it hires the investors or brokers instead who work for DHA. DHA offers 2 kanal (0.10 ha) of land for 8 kanals (0.405 ha) of land acquired. It means after giving away 0.405 ha, DHA in exchange provides 0.10 ha and keeps other 0.30 ha for itself to build roads in the area. The importance of road building could be imagined from the excerpt below:

It was an incident that showed another side of land acquisition, that did not need to be the farmland, but houses too. It occurred while I was in the beginning of the research and roaming around in Dera Chahal in 2014 and found a group of elderly men sitting outside a small hen's slaughterhouse. I could hear their laughter from afar which caught my attention in the first place. There were small, tiny houses on the right-hand side of a narrow road and the shop was across the street surrounded by agricultural lands. I introduced myself as a researcher and expressed my desire to have few minutes from them, they started dispersing in no time and that left me in an entirely mind-boggling situation. I bought myself some time to regain the confidence and finally mustered up the courage to talk to the man from the group who stayed back. 'What do you want? He asked angrily. 'I already have received the notice to abandon my house, now stop bothering!', the man yelled. He was crossed because he thought of me as someone from DHA, demanding him to leave his land. DHA has been the most powerful developer, used to purchase land from villages for the construction of gated communities but the old villager's reaction conveyed much more. Later when grey-haired realised that I am just a researcher and have got nothing to do with DHA or his land, he became a cooperative and hospitable being. He showed me his home and said, 'DHA wants us to vacant the house as they need to demolish it to widen the road'.

The portrayal of land acquisition by DHA seems fair, but it is not only intricate but a far cry from the ground reality, coming from gamut of people's experiences. So, DHA issues files—one file (each of 0.05 ha) for 0.405 ha costs 55–60 lakhs (17,693–19,301 €[5]; Researcher's data, 2013–2015), and now, the price for same amount of land has spiked up to 13 million (42,971 €), (Bhangali, 2021). The land in *Harpalke* was sold in 2004; at that time, 0.405 ha of land was for 11.448 €, and the same amount of land was sold in 2018 for around 261.660 € (Harpalke, 2018). However, there is another party involved that acts as a bridge between DHA and the landowner who is known as broker/investor or middleman. The investors who actually work on behalf of DHA are usually trusted members of the community and help fix deals at lowest possible rate through bargaining and gradually upping the rates with respect to flexibility or rigidity of the holder; variation in prices also depends

[5]The PKR conversion to euros is according to current exchange rate on 12 May 2023, i.e. 1€ = 310,86 PKR.

upon the location as well. This truth surfaced again and again over the course of the research, whispered by brokers, affirmed by landowners, and echoed by villagers. All pointed to the same unsettling reality: DHA relies on local insiders to penetrate communities and carry out its work, turning neighbours into intermediaries and blurring the line between collaboration and coercion. A DHA broker from *Chachuwali* zeroes in on the fact that even if the brokers are from the outside, yet they need to involve someone from within who is fully informed about landholdings and people and is also trustworthy for the locals (FM-Revenue Office, 2021). However, land prices have registered a steep increase with time. *Chachuwali* resident (2021) had 1 acre (0.405 ha), out of which 0.202 ha was sold for 50–55 lakh PKR (16,328–17,961 €) in 2004, whereas the other half of land was sold by the owner in 2013 for 2.5 crore PKR (81,644 €). According to a villager (Kalasmari, 2022), 1 acre (0.405 ha) of land was sold for PKR 2,300,000 (965.07 €) in 2003, and now, 1 marla (25.3 m^2) costs 257,353 € (FM-Kalasmari, 2018; 2022). There are cases where people sold marla (25.3 m^2) for 3,500,000 PKR (11,259 €) reported by focus group discussion (2021). Currently, 1 kanal (0.05 ha) land is of 35–50 million PKR (Gujjar-Chachuwali, 2022).

Rates have risen and fallen in step with DHA's shifting desperation for land, revealing how negotiation unfolds within an unequal field of power. When DHA urgently needs to acquire a particular parcel, the balance tilts slightly toward villagers, allowing them to leverage their land and negotiate for better terms. Yet such temporary advantages do not alter the underlying extractive logic. What appears as negotiation remains embedded in broader structures of dispossession in which the powerful ultimately set the terms of exchange. In this terrain, the price of land becomes a measure of how effectively residents can resist, foot-drag, or complicate acquisition. Their resilience and strategic bargaining may momentarily slow the machinery of extraction, but the deeper power asymmetry remains firmly intact. Although not every village got an opportunity to decide for themselves, the general pattern for land transaction has been consistent wherein investors play a vital role in making a deal with owners (without bringing DHA to the forefront and acting as a veil that keeps DHA hidden until it wants to be no longer secret) and getting the price fixed, and owners are being paid by the investor himself, and files are transferred in the name of investor who becomes the owner of land being offered by DHA. He is then free to sell it anywhere he wants. So, investors provide land to DHA by purchasing it from the landowner, and he himself becomes the owner of the land as a personal purchase given as compensation for the land acquired. The investor directly pays the landowner in cash, and file (entitlement to land) goes to investor who is now the landlord, and DHA gets the land for real-estate purpose. People from the village who worked as investors have become billionaires by exploiting the community's trust. DHA sends its investors to do all the work for them. '*This village had an investor, he started scamming people, otherwise village would have been vacated long ago. He did so many things, he took the land, sold it to DHA, used people's money, that pushed people 6-12 months backwards. See, if you buy 100 acres (40 ha) and make payment to sellers after a year, then it is better to go to DHA directly. He went a lot ahead in his life. Right now, he is owner of 35 Murabba (354 ha) land within this Lahore city and the places he owns on the main are a separate story, his greed is still not satisfied, such is the addiction of money'* (Chak Dhira, 2022). However, apart from *Charrar, Pathanwala* was another village having 10 homes (families from the same lineage) with 10 *Murabba* (101.17 ha) of land where DHA directly dealt with people back in 2013, that is when the village was sold (Pathanwala, 2018).

People also call DHA investors goons or the land mafia, who can go to any length to get the job done (Lawyer-Durgpura, 2022; FM-Durgpura, 2021). Gujjar (Chachuwali, 2021) shared an incident of someone who he knew well, which speaks volumes about the role of investors. *'DHA has its investors and has given them all the liberty. DHA never comes to the front, even when you go to their office, they will tell you to wait and then they will send a person on their behalf who is the broker/investor. Four months ago, incident happened to a well-established and a resourceful Kamboh. He had a well-maintained land, with guava and sweet lime orchards, peacocks, hens, goats. He was not ready to sell his land unless DHA paid him the desired amount, when things did not workout. DHA asked its investor to get the land by any means for which they paid him 50 crore (1.634.758 €) to make this happen by hook or by crook. Investor distributed 5 crore (163.500 €) amongst the police officials including S.H.O and his deputy. An order was issued for a raid in the middle of the night. He brought 40-50 policemen, 10-15 other persons, accompanied by D.C. They also brought tractors and bulldozers. In the blink of an eye, the walls were bulldozed, entire land was ruined, the animals looted and nothing saved. He was not paid even a single rupee as compensation. He could not file a case in the name of DHA either, because when he would be asked, who did this? He could tell the name of a person (investor), the accused investor might work for DHA yet was not a DHA employee. Roads have already been built on that area'.*

Some people think DHA only acquires agricultural land and leaves the settlement as it is; thus, inhabited area doesn't get displaced in the process. However, data collected from people (2014–2022 and personal observations) prove otherwise. Some part of *Haveliyan* village was intact in 2021 but was completely removed in 2022, and *Heera Singh* is nowhere to be found except for its shrunken graveyard. *Sangatpura* is history, and *Gobindpura* has disappeared into thin air. *Chak Dhira* is on the verge of removal; *Natha Singh* is another such example where the entire village has been sold except for one house whose owners are not ready to leave until DHA comes to terms with them. *Chachuwali* residents (2021) are certain that like other villages, this settlement will also be removed in just a matter of time. Therefore, the final fate of a village is totally in the hands of DHA, according to realtor's needs and will. *Chak Bhart* has been pocketed, and *Bhangali, Kalasmari* and *Harpalke* are also walled off. *Charrar* and *Durgpura* are amongst settlements which are service sectors for the residents of the gated communities.

DHA does it secretively to avoid people having time to organise to resist. The other benefit for hiring investors to do work on behalf of DHA is a strategic move to keep prices as low as possible. As DHA is a brand in housing, buyers tend to increase their demand the moment they come to know that this land is going to be taken by DHA. *'When investors started buying land (around 2002 to 2003) from the village, no one knew at the time that it was basically DHA, people were unaware that DHA was coming because we could only see the investors. Broker was offering 8 lacs (2573 €)/ 0.405 ha, whereas we demanded 10 lacs (3216 €). After a few years when DHA finally came out of the closet, deal was sealed at 20 lacs (6434 €) per hectare. By 2003 agricultural land of the village was completely sold. Land was rented out by DHA to tenants for farming. In 2007-2008, topography took a new shape'* (Chak Bhart, 2022). Similarly, in *Heera Singh*, the locals said that they were used to army conducting military exercises in the area as they often do, given that it is a border area. Only after they had sold their land did they learn that the land was being surveyed for DHA development. In 2002, most people had sold the agricultural land for 17 lakhs (5468 €)–20 lakhs (6434 €) per 0.405 ha, and those who sold in 2007, prices went five times higher the earlier price, i.e. ten million PKR (32,169 €) per 0.405 ha. Later in 2015, about a dozen households who waited out the longest sold their residential land for 42 lakhs (1351 €) per 25.3 m^2, an astronomical price for land matched only by land prices in the highly commercialised properties in the inner city of Lahore.

The local land assemblers are effective in persuading the whole community to sell. They also are able to sell all the commons, and the compensation is divided amongst the villagers. Typically, the most are gained by outside investors who buy the land in bulk and then sell it to DHA. The real gains are made by the large investors who have both holding power and inside knowledge of future development plans in the area and finally DHA which sells the individual developed plots at many times the original value of land. The linkage with the army makes them more attractive for private investment as there is less fraud than civilian-run schemes. However, in many places, the villagers complained that the middleman never even informed the locals that the land is being acquired for DHA. After taking over the agricultural land, once they erect a boundary wall as a symbol of their ownership and colonial supremacy, locals are no longer allowed to farm the land and also forced to leave the land fallow to let it die on its own pace if needed (accounts gathered from researcher's own experience through MPhil thesis on *Shuhada* Town-a DHA's project, 2013–2015).

The most profits are made by large investors who got large parcels of land through local assemblers and also invested in buying files for future plots. The villagers who owned no agricultural land got the worst deal as their typical homes of 10 *marlas* (253 m^2) or less were of no interest to DHA, which does not buy settled land except in special cases. The best they can do is to rent it out to migrant workers but that too has few takers. The poorest and often illiterate were also cheated by corrupt dealers and land assemblers. Many of the farmers who sold their land and moved to far off places were also cheated, attacked and harassed by powerful locals in places far away such as *Layyah* and *Pakpattan*. Only those who had local connections or relatives were able to prosper in the new places (Patwarkhana, 2016; Harpalke, 2018; Fi-Kalasmari, 2018). The division in the land did not turn out to be in favour of those who had small landholdings and several shareholders. The money came from a land transaction made blood relations to stand against each other. *'DHA has severed several blood ties'* (Natha Singh, 2021). However, in villages who are on the verge of removal, *'People who only had land over which their house was built, looked at DHA deal with hope as many members were living in a crammed space, could buy land elsewhere and could build walls around it'* (Chak Dhira, 2022).

5.3.3 Soil Acquisition

Landlords are running businesses even by selling the soil, (Natha Singh, 2021).

The soil acquisition process (see Fig. 5.18) is another way to compel people to leave their farms to facilitate high-income faction of society. It is a simple process both in terms of acquiring earth to make the raised base of asphalt roads and also very straight-forward exercise to prepare land for transaction down the line. It is an undertaking in which compensation amount is fixed, i.e. 2 lakh PKR (618 €[6]) for 1.5 m top soil/acre (0.405 ha) between landowner and contractor hired by Frontier Works Organization (FWO)—an engineering branch of Pakistan Army. Land remains under the possession of landowner, but soil is sold to the contractor. The extracted fertile top soil is used to raise the road level by 0.9 m on average as shown in Fig. 5.19 (Villagers 2014–2022).

When a farmer sells the top soil which is the soul of agricultural land, he either has to wait for 3 years to allow land to be resurrected or has to spend more on chemical inputs and fertilisers to add. However, farmers reported that once the top soil is gone, land's ability to produce can't be restored to previous level (Kohriyan,

Fig. 5.18 Selling 1.5 m of top soil to pay off debt or marriage or medical expense cripples the ages land

[6] According to current exchange rate on 13 May 2023, i.e. 1 € = 323,40 PKR. Since currency has been changing very frequently, I am mentioning the rates in Pakistan rupees and euros to make it fathomable.

Fig. 5.19 0.9 m raised road—at the cost of productive soil and land, 2021 (The image was published in other places before (Javaid 2023, 2025))

2016; others 2014–2022). The small 'compensation' may help cash-poor farmers to meet immediate needs without much ado, but the environmental and ecological damage is only externalised knowing the term 'compensation money'. It may not be incorrect to say that selling top soil for a few bucks is killing the goose that lays the golden eggs. It was reported at several instances by the interlocutors (2014–2022) that '*Once the top soil has been removed, the land can never be the same again, either it needs to be left for years to gain its strength back or increase the input of chemical fertilizers substantially to enable the land to gain the same productivity*'.

Ecology has been degraded in multiple ways, other than making the land unfit for crop production; once the land has been lowered, these agricultural lands have been used as waste disposal sites by the municipal waste disposal staff. In village *Mandianwala*, a few hectare-sized ditch created by soil mining was used as a garbage dump (see Fig. 5.20) by the municipal waste disposal staff who managed to save fuel by paying a small price to the local landlord who was keen to have the land raised. Later, the same landlord sold the garbage to others who needed material for filling. Similarly, abandoned village ponds in *Rakh Padri* have become a garbage disposal sites without any official sanction or planning. *Phuller Wan* also had a waste disposal site; currently, none of them is operational (Lawyer, 2022).

The desperation of farmers for cash for various urgent needs is leading them to destroy their own productive assets. The need for cash will only keep increasing even when they have sold their land and have moved to an urbanised lifestyle in an urban location.

Fig. 5.20 Waste pit in a village called Mandianwala, 2016

5.3.4 Graveyards: Another Site for Land Acquisition

> *DHA asked us to sell the graveyard. Our parents are buried there, tomorrow we will have to have a place to be buried, then our children will be back to the same graveyard. Selling graveyards is equivalent to selling one's parents and who would sell the parents, what would we leave behind* (Fi-Pathanwala, 2018).

Peripheral land is money-engendering machine for DHA; therefore, it brazenly wants everything that exists on the surface; it is put up for 'sale' including all kinds of structures such as schools, roads, graveyards, shrines, drains and wastewater ponds (Villagers, 2016–2022). The other popular site for land grab is burial ground (see Fig. 5.21) as dead can't resist being displaced and disrespected, while the ones still breathing are in dire need of cash. A woman (Heera singh, 2016) expressed her disbelief and remorse on newfound mindset of people, *'Did anyone ever hear of such thing?! Did anyone ever disgrace a grave?! People have sold graves in greed…the only major concern is now Money!'* (FW-Heera Singh, 2016). In a house full of *patwaris* (revenue record holders), brokers (2016) candidly told me, *'After selling land people sold graves/ graveyards and shrines (to DHA). One marla (25.3 m^2) of the graveyard was sold for PKR 10 million (33184 €) especially if grave or graveyards is at a good location or comes in a commercial area. Now every person who used to plough fields is now a property dealer. Once he earned 2 lacs PKR (645.16 €) from one deal, he made it into his business'* (FM-Patwarkhana, 2016).

However, most of the altercations that led to casualties happened for defending place and space after death. The grave transaction was simple where the only condition for compensation was a documentary proof with the deceased while unmarked graves were shifted without a compensation. The maximum bid being heard so far for selling a grave to DHA is PKR 10 lakhs (3318 €) (Villagers, 2022). This

Fig. 5.21 'Infamous' graveyard of Heera Singh: 'This graveyard is not a DHA's property but Heera Singh' (left). However, over the last decades, the graveyard has shrunken (right)

phenomenon is still not as rampant because people still are culturally and emotionally tied to place not only because of their departed loved ones, also because they want to secure a piece of land for their own burial which otherwise has become a costly business. Graveyards or shrines have served multiple purposes for rural dwellers: as open common space, a spot to get together, annual festivals and grazing land for livestock. Till recently, the annual *melas* (festivals) at various shrines dotting the local graveyards were celebrated at many villages. Each village graveyard was a living history of the community and literally rooted in the soil. The land acquisition has not only intimidated people by the idea of losing farmland or home but also made them insecure regarding the availability of land for burial, as old couple who shifted to gated community from a village which was obliterated said, '*We both (couple) were thinking that where will we be buried; it's all about fate wherever it takes us. But we would have been easily buried if we were there*' (Fi-Guldasht town, 2016).

Its effect multiplies when it happens to minority or a non-Muslim sect. It works well for divide and rule in a community. The fact that it is persons from the same sect who make the transactions, such occurrence also reveals the power of cash in winning over, unfolds the dark side of trust and network over which communities thrive and survive. A Christian lawyer from *Durgpura* (who is also handling the local case of his community's graveyard) said, *'We won't hesitate if we have to fight legally to protect our graveyard which DHA is dying to acquire. They have not even touched Muslim graveyard but they are after ours. People themselves are more responsible who become accomplices to DHA for a chunk of money. DHA wants to shift Christians' boneyard to make it a part of military dairy-farm so that DHA can use the property as per their wish. Other unfortunate thing is that usually people get divided into groups in such matter which makes even more difficult to work things out. Historically this cemetery was known as Mariyanwala (a place where Sikhs used to burn their loved ones to ashes). Once they left, people in the vicinity started using it. This is a combined graveyard for Christians from Chak Bhart and Durgpura, this parcel of land was given by people of Chak Bhart to their servants. My grandfather was the first to be buried in this God's acre in 1980'* (Fi/22-Durgpura, 2022).

Monetary compensation was irresistible for Charrar residents, but nevertheless graveyard became the site for resistance. The residents of Charrar fought against the DHA to save their graveyard. In Charrar, the villagers had banned the burial of non-locals in the village graveyard. Given the astronomical densification of Charrar, the villagers were afraid of losing space in their ancestral graveyard. They had a practice of collecting money for indigent migrants and sending their remains to their places of origin. The Christians in Charrar too have a large well-maintained graveyard. People in *Kalasmari* took a stand to save the graveyard. They have had it since 1965; most of the rural land went underneath Green City Malls already, but then people stood up to save the remaining area (Fi-Kalasmari, 2018). *Durgpura* is another village where the Christian community had to protest to save its graveyard. In a village where Muslims and Christians have been living together for ages in harmony, the Christian community's struggle to secure their graveyard inculcated a sense that a minority is being treated differently as DHA had not touched the Muslim graveyard. The cultural life of the villagers and their identity was tied to the landscape. The festivals, on the one hand, are markers of collective identity for villages; on the other, they have long offered moments of joy, renewal, and emotional return for married women. When women come back to their natal homes for these occasions, the village shifts almost imperceptibly—doors stay open a little longer, courtyards fill with familiar laughter, and neighbours slip in and out as if time has folded back on itself. No formal invitation is needed. Their presence during festivals is understood, expected, and absorbed into the very rhythm of village life, into the seasonal pulse of celebration, kinship, and memory. These homecomings momentarily knit together what distance, marriage, and urbanisation have stretched apart, reminding everyone of the ties that endure despite the worlds that change around them.

Village graveyards are not merely burial spaces; they are grounds of identity and quiet, enduring resistance. Many villages that once kept their graveyards open and unwalled—spaces that were fluid, communal, and accessible—have now begun to enclose them with walls, marking a shift toward protection, assertion, and the need to safeguard what remains of the village's collective inheritance. Local leaders realising the sensitivity and pride people feel towards their graveyards, too, have taken lead to raise funds to get the graveyards walled and build *Janazgahs* or covered spaces for the burial prayers. A local resident, who also works for DHA, told me, '*I myself bought land (in Chachuwali) and dedicated it for the graveyard* (see Fig. 5.22)

Fig. 5.22 Open Janazgah (right) and graveyard (left), Chachuwali, 2021

and I am also building a spacious Janazgah in there' (FM-Revenue Office, 2021). It is now a new landscape with hardened boundaries and identity politics. While a nostalgia for the lost land remains, it is still harder to recreate the lost social landscape.

Villagers recounted a cycle of events throughout the year and across the landscape. But more importantly, each village graveyard was a rooted living history of the community. The land acquisition has not only intimidated people by the idea of losing farm land or homes but also made them insecure regarding the availability of land beside their departed loved ones. It was mentioned time and time again that the members who have migrated to other locales still wish to be laid to rest in their own village graveyards. The local communities still look after the cemeteries collectively. Nevertheless, graves and people attitude towards them also depict change as how quickly money can switch a person from fighting to protect a grave to sealing deals: from 'a collection of heritage and family legacy' to 'just bones that can either be shifted or be forgotten' for money (see Fig. 5.23). Although this particular transformation is not as rampant as fire and villagers have fought to secure the burial grounds and to protect shrines that are normally found in all villages (see Fig. 5.24), I have heard people talking about graves as after-life home that should also be spacious and also drawing comparisons between city and rural 'post-death-living, *'When a man dies, he used to get a big palatial grave, now it is available in size as it is in cities. If you really want to see what a big house looks like, then go and gander at grave, it is roomy and strong. If you go to Lahore, they twist the bodies and bury it in a small hole-sized place'* (Bhangali, 2021).

Fig. 5.23 Graves surrounded by rubble

Fig. 5.24 Shrine linked with Dahoorwala

5.3.5 Environmentally Stable Zones at Disposal

Aggressive land acquisition for development is leading to land-use change and tangible environmental degradation in terms of piling of waste, dumping of natural resources, dust, smell in the air and contamination of groundwater and allergies due to dust, amongst the physical manifestation of change. Developers keep the gated communities clean by dumping waste in small settlements, treating them as elite's trash bins. '*They don't let us throw trash outside, if we do, they file a report against us'*, an old woman from *Chachuwali* articulated the unjust treatment towards those on whose 'wealth' DHA had been profiting (FW/1-Chachuwali, 2021). What she told me was validated by a warning from DHA, written on the wall of an enclosure by them (see Figs. 5.25 and 5.26).

The extremely useful cow dung has also turned into waste. It cannot be used as manure because agricultural land has disappeared, and as for its use as fuel, there is hardly any place for drying dung cakes. The educated females are averse to taking care of animals, which was also associated with collecting cow dung, making dung cakes, applying them on the walls of mud houses for cooling affect and keeping them as source of heating while cooking. So, cow dung and dung cakes have become commodities which are saleable products on Amazon. Food grown using cow dungs as fertiliser is advertised as organic and as such has gained higher commercial value. Food grown from such natural fertilisers is expensive. This is the irony of 'development', as one's gain is someone's loss. The increase in the use of packaged products and plastics has also created a crisis of disposal. Given the loss of commons, roads

Fig. 5.25 Warning on the 'mighty' wall: 'One who dumps waste in DHA premises will be handed over to the police; DHA Lahore', 2021

Fig. 5.26 Gifts of urbanism—trash and solid waste. Residential community (left) and trees are buried in trash (right)

Fig. 5.27 Face of destabilised ecology

Fig. 5.28 'Aesthetic living' is a privilege, Bhangali, 2021

have become waste dumps and the villages are full of garbage in all parcels of land not in private hands (see Fig. 5.27).

The groundwater contamination is another ecological distress that has struck every village surrounded by gated communities. It happens as septic tanks contaminated the groundwater; people in villages also have started building modern sanitation system that contaminates groundwater, yet it is not common in villages. The disrupted fresh water channels giving foul odour create stagnant, black water ponds, banning the entry of people who once belonged to that land to ensure 'well-being' of the rich. Furthermore, mobile towers that must not be installed even in the proximity can be seen in the middle of a village such as *Bhangali, Kohriyan, Kalasmari* and *Durgpura,* which shows the manifestation of callousness of DHA towards village communities.

DHA can be considered the harbinger of large-scale development in the area, there had been an on-going slow appropriation of land in the suburban villages by individuals and businesses took advantage of the rent gap and set up their operations in the area.[7] Unlike activities owned or controlled by members of the community such as small crafts, milk and vegetable production or plain residence for commuting workers, more gentrifying trends in the area had set in as well such as the use as a site for waste disposal (see Fig. 5.28) while keeping the high-income localities clean.

[7] Since the 1990s, farmhouses (luxury mansions in the countryside) became popular with the elite in Lahore. Inexpensive land was also acquired for setting up small-scale operations such as furniture-making workshops, gypsum tile making, waste recycling, gaming facilities and schools, etc.

5.4 Tricks, Deceit, Power Play to Acquire Land

In the village *Lehna Singh*, a wall was built around the village after the agricultural land was acquired, blocking the residents' movement from the village (FM-Lehna Singh, 2018; 2016). Villagers from *Harpalke* and *Natha Singh* raised the issue of on-purpose non-provision of basic amenities such as natural gas as one of their tactics (FM-Harpalke, 2018; Natha Singh, 2021). All common lands, roads, government schools and graveyards are also acquired in a process, promoting greed while appearing to deliver value in monetary terms. Usually, the common lands under state control are all parcelled, and their compensation is divided amongst the villagers. In *Heera Singh,* some members of the community put up strong resistance against a surreptitious appropriation of the school land by a land grabber in cahoots with the education department. *Heera Singh* is one of the few villages where even homes have been sold, and the village is obliterated except for two houses and a mosque. The last group of homes which had negotiated a good price through a tenacious show of unity before moving were given an unrealistic target of two days to move their homes. They managed to fulfil this condition through a remarkable organisation of the evacuation. However, they narrated this as a tactic used to force the villagers to agree to a lower rate (Fi-Guldasht Town, 2016).

A DHA broker who belonged to the village *Lehna Singh* narrated the ordeal of injustice and resistance. The people (of *Lehna Singh)* had filed a case against the disruption of the power supply as DHA was removing the overhead wire network of poles from the land around the village which was now DHA property. DHA surreptitiously removed all the poles in a single night and connected them to the underground power supply of DHA. The villagers no longer had the will to continue the case. They already had access to electricity, and the physical removal of the poles - once seen as a potential obstacle to DHA's development work and a bargaining chip for demanding additional amenities - had eliminated their last leverage. They had earlier successfully negotiated to get their graveyard walled so that it was safe from future encroachment. DHA had also provided them an elementary sewerage system as the drainage ponds around the village were no more, and there was no passage for waste water. However, they complained that the system provided by DHA was of very poor quality. Many villagers had left the area as it was not pleasant to live there (Lehna Singh, 2016). People in *Chachuwali* also expressed their anger due to sewerage problem as DHA disrupted the passage of water that used to go directly to the sewerage or wastewater pond (see Fig. 5.29) due to which people living close to wall complained about foul odour. Also, DHA provided two out of three phases of electricity, which caused the issues of short circuit and damage of electric appliances (M; FW/1-Chachuwali, 2021).

Many poor landless workers found commuting difficult as DHA often stopped young males and harassed them. Almost a fourth of the village houses were empty. During the focus group discussions at *Haveli* (a place for livestock-keeping), a young man expressed, *'One of the main routes were blocked by DHA, they do allow bikes or cars to pass but they do not let the trolley loaded with fodder to pass'*. He

Fig. 5.29 Wall separating wastewater pond from the village, 2021

Fig. 5.30 'New colonisation' developing 'detention camps', Durgpura, 2022

mentioned repeatedly that *'DHA has made lives miserable; we can't take the animal out in wastewater pond, it is difficult for animals also in summer, people are forced to sell their animals which are the livelihood'* (FM-Chak Dhira, 2022).

The land that has already been taken is enclosed by the DHA for the building of gated housing that can be seen from afar due to symbolic walls (see Fig. 5.30) which realtors erect with multiple literal and figurative meanings. Interlocutors, throughout the years of fieldwork, belonging to different villages reported struggles they had to face due to enclosure cum exclusion. Keeping cattle for domestic is a normal practice in villages whether they are into farming, milk business or not. Several dwellers had to sell their animal because they were not allowed to take their flocks out from the walls erected by DHA; otherwise, they were charged with heavy fines. Animals are undesirable entities which in most areas are not allowed to cross the gate, and taking animals for grazing in left-over land comes with heavy penalty, *'Beyond the wall, there were open tracts of land as far as eye could see, we used to leave animals there for grazing, fodder was in abundance and for free. Can you see those bungalows? Heaps of wheat used to be stacked up on the land where there are big houses now. And, where you are standing right now, this entire left-side of pathway was crowded with cows and sheep. If you had come 10-15 years ago, you also could have seen variety of vegetables which the villagers used to grow for personal use. Now, this place is no longer the same. They (DHA) do not allow us to take animals outside, if we do, they penalize us with PKR 5000-7000 (16.5-23 €), this is why people are left with no choice except sell their livestock'* (W1; FM-Jindra Kalan, 2022).

It is interesting to see that people's livelihood and their professions create a difference of opinion towards the same issue as they tend to have different stakes. The educated faction of a settlement especially male members from age group of middle teens to mid 20 s and also the established property dealers (2021) expressed a different viewpoint in response to DHA's restriction on villagers for taking their animals out, which is nothing less than seeing themselves from the eyes of elite. According to them, DHA is right in doing so as who would want to live on dirty roads, and they are just taking measures to keep their locality clean. In fact, a property dealer and investor from a village looked down upon living in a place where he was born and raised and also tried to convince fellow community members to move to a better place leaving filth behind *(gund se niklo).* In traditional rural living, where land is deemed as 'mother', a pet animal is like a 'child' to them.

During a discussion with men in *Jindra Kalan* (2022), an elderly woman joined our conversation, and group of men greeted her with utmost respect and introduced her as *'Amma ji'* (mother—title of respect for an elderly woman) and said, *'These are her animals'*, to which Amma ji replied with remorse, *'Child, I used to have a lot of cattle, we were into milk selling business which was our main source of income and supported me throughout, when DHA came here, I had to sell them, these are the only ones left'*. She was an open-minded, progressive rural woman, who made sure that her children would get education regardless of gender and also helped her daughter-in-law to complete her studies successfully. She extensively talked about the rural life and perks without covering up. *'If I compare the old and new life, the former living was way better, we had less money but it never felt like that we had nothing. We had healthy food, quality relationships, open land, clean air, water and whatnot. This "change" has ruined everything'*. Figure 5.31 screams about the unsaid, such as lives at two extreme ends, a new life across the wall yet hard to climb.

However, young girls/young mothers and some middle-aged women (30 to mid 40 s) had a different complaint towards building of walls. They take it as their journey from freedom to confinement. The detention walls have denied their freedom to roam freely in the premises of village which have already been encroached by the

Fig. 5.31 A colonial wall where two worlds divide, 2022

DHA. Aged women have advantage of their age, and others can't put restrictions on them, but young girls and middle-aged women have to be very cautious, and majority of them prefer to stay inside than to be questioned by men. As women in *Chachuwali,* from three different age groups, gathered at one place and expressed their perturbance during different encounters, '*In old times if boys and girls were* sitting *together, no one raised eyebrows because it was not an issue at all. Now if we (girls), by any chance look outside from a door then brothers would instantly question what are we staring at. Things have become tougher for us (young women and girls). Now we are confined to our homes, we just don't go out. This is one of the main reasons why even women get-togethers don't happen very often'* (FW/2-Chachuwali, 2021).

This is not just a wall of partition that speaks for the change, but road building itself is a strategic coercion for giving up land. The road construction caused damage by the complete land-use change on the one hand and by taking away the productivity of arable land from where the top soil was excavated for the base of new roads on the other hand. The soil excavation is socio-ecological degradation of arable lands in areas outside DHA and has a strategic role in persuading people to sell their land. The scrapping of top soil causes unevenness in landscape by deepening the ground level, creating ponds of water after rains as well as obstructing the flow of water courses and irrigation channels. So, if one person decides to sell the soil, it instils fear among neighbours about loss of the quality of land, agency to sell and the power of negotiation over money when it comes down to this. The unevenness in land has created pits where water gets filled in (see Fig. 5.32).

Fig. 5.32 Perils of 'uneven development': Outside of the Christian's graveyard, 2022 (left); 'Hollow' development 'hurts deep'—holes in the heart of earth (right)

5.4.1 Coercion, Violence and Resistance

Villagers reported that hired 'staff' are kept only for intimidation purposes, and another group exist that could terrorise through life-threatening scuffles. '*DHA is a mafia, they are no different from goons*', this is how villagers in certain villages name them. DHA's manner of dealing varies with the owner's landholding, the strength of the community and the social class they belong to. DHA's injustice, violence and desperation for taking over land increase against the weak and the marginalised. Resistance is a privilege that is only permitted to the strong. The marginalised faction often goes unheard and unseen. DHA just treats the land as vacant and starts cutting roads over those available plots without even asking the owners. There were a number of cases in *Kamaha* and villages around *Baba Shah Noor*, where owners have proper paperwork of their plots, and still none of them got the compensation for using their properties without asking. DHA does not hesitate in putting people behind the bar (FM-Durgpura, 2021).

According to villagers, the prices they were offered by DHA proved very effective, and the villagers agreed to exchange their land for a price they thought they could not refuse. In *Charrar,* they emphasised the unity of the community and how even the minority Christians of the village were part of the committee through which they negotiated rights for the village from DHA. Their relationship with DHA was largely amiable, and they managed to get most of their demands fulfilled. *Charrar* now has multiple pathways connecting it to the outside, along with access to sewerage, water supply, and gas. These amenities did not arrive by default but emerged through sustained resistance and subsequent negotiation with DHA.

Village communities find their strength in unity, and therefore, there were instances where tables were turned due to resistance being put by the villagers. For instance, people in *Chak Dhira* stood against the DHA in every possible way be it through physical showdown reluctance to bow down or verbal representation of concerns, apprehensions and demands, and dwellers succeeded in saving their main official/government route (*sarkaari rasta*) and got sewerage facility too (Chak

Dhira, 2022). Also, Gujjar in *Chachuwali* (2022) highlighted the friction between DHA and people, '*Chachuwali denizens were completely against selling their main government route to DHA, they used to say that as long as their people were sitting here, they would keep their govt. route at all costs. It was a legal battle which they fought and had stay on route for 10 years. Afterwards, 10-12 influential, clever and stronger members of the community shook hands with DHA and sealed the deal for PKR 20 million (66301 €). Rest of the village could only cry in disbelief but only too late*'. There have also been instances of violent conflict such as the one in 2007 when one person died during protests against DHA's attempt to unlawfully occupy part of the village graveyard.

Rural dwellers may tend to sell their lands 'by choice', but their resistance for keeping homes and graveyards out of the reach of market speaks for their deep emotional attachment to their soils and cultures. A group of men used to stay at graveyard to keep an eye on land grabbers in *Durgpura* (L-Durgpura, 2022), and even women came out of their houses and blocked the road for three to four hours to make themselves heard (Fi/22-Durgpura, 2022). Similar way of guarding was seen at *Bhangali* at holy shrine (Bhangali, 2021). Another brawl happened on DHA's road construction at *Harpalke* village, resulting in the arrest of 15 protestors (Dawn 2021). *Barki* village reported fight over the main route when three young men were the face of resistance, and eventually, one died in the hands of DHA (FM-Durgpura, 2021).

5.5 Peri-Urban Areas: Site 'of' Social Degradation

This section aims at communicating people's loss and gain narrative that is intangible and has been experienced by the respondents as representatives of the communities, regardless of their social class, monetary wealth or assets in the form of number or type of animals. They included people ranging from big landholders to the majority small landholders to the landless and wage workers or those on monthly salary or running rented shops. It incorporates men and women from three different generations. The purpose of having range is to avoid biases and to grasp nuances. Above all, it creates a rooted narrative of loss and gain from their mundane life experiences and routines. It begins with the most obvious changes such as in food and water and intangibles and includes expression of the even more powerful intangible determinants of their life through people's own voices.

5.5.1 Degraded Food, Water and Health Quality

'People have got huge amounts of money, in millions and uncountable sums, now sellers have easier lives than before, yet this is not development. Let me tell you what development used to look like, a university where you were a student, from that

point till here, there were guava orchards, mango tress looked nothing less than a Bore tree, wheat was all around, water-well was also a get-together spot for members, that was life' (Bhangali, 2021). Another interlocutor who was an old teacher once said, *'This is not a way-forward, it's a lie who says so! We are getting destroyed and moving backward'*, (Chachuwali, 2021), who bought 8 ha elsewhere after selling his 0.40 ha to the DHA and had a large beautiful house in the *Chachuwali*. He also lamented what people were getting in exchange of land for enticing compensations: *'The quality of food has been tremendously affected since we have lost our means to produce food for domestic purpose only, now we are buying everything from the stores and everything is debased or artificial, water has been contaminated due to which hepatitis and jaundice are now common diseases. Now, DHA has carpeted the roads but the road construction is still going on all around the settlement, until that point 90 persons out of 100 have contracted some form of asthma allergy to dust pollution, breathing-related issues are not uncommon anymore'*.

The above-mentioned excerpts are taken from direct conversations with interlocutors in their respective villages, but they were all subsistence farmers at one time. They had relied on nature and its resources and were convinced that nature offered a stability that modern farming is incapable of. It was also intriguing to know that villagers clearly drew a line between getting rich and well-being. Denizens realise and don't deny the increasing need for cash but more so because modern farming has destroyed the soil which was able to feed, sustain and meet the basic needs of their lives. Water wells closed down (see Fig. 5.33) with the arrival of development, ground water is unfit and now access to clean water is one of the issues.

A well-established agriculturist and landlord explained, *'Every household had animals generally 2 to 5 as source of fresh milk. Food was in abundance especially and people were very generous. If someone's animal in a village had dried and he asked for milk, people would feel pleasure in helping him and neighbours gave milk in large steel vessels. And, elderly women of the household would add a chunk of butter in a glass of buttermilk if someone had come to the door for buttermilk, out*

Fig. 5.33 A closed water well in Chachuwali which was operational till 2006–2007, 2021

of concern for the person's children. So that, he could be able to serve 'pratha' (a fried, layered flat-bread with ghee inside) a dry one' (Chak Bhart, 2022).

It is no longer in people's hands to be generous when they don't have for themselves. This is one proof that places don't change alone but mutate even the spirit. When people had abundance of food, so were their hearts that believed in sharing and magnanimity. Now, it has become difficult for the majority to be hospitable. As a woman who migrated from *Mananwala* to *Mehar town* and works as a maid (2022) expressed, *'Since it has become expensive to be a good host, now it depends on the type and position of the guest, according to which we decide upon the type of food or snack that we are going to serve, rusk or cake'*. Of course, these peri-urban villages were never alienated from cash-based exchanges, yet they largely remained independent of the market's grip and the tightening noose of money for meeting basic needs. Women carried buckets of cow-dung cakes on their heads to sell in *R.A. Bazar* or *Bhatta Chowk*; families sold milk; others took fruit to nearby *mandis* for sale (Fi-Chachuwali, 2021). Even with this circulation of cash, their lives did not revolve around money-making, nor were they caught in the labyrinth of seeking ways to multiply it. As recently as fifteen to twenty years ago - before the urban model tightened its hold - villagers lived at a different tempo, outside the whirlpool that has swept city dwellers into a relentless rat race. The fast-paced pursuit of an illusory happiness tied solely to income and consumption had not yet seeped into their everyday worlds. It is not too long ago, when city was divided in terms of not only landscape but also the way of life which citizens were living, which was different from those at the periphery and rural. *'Just like landlords, even landless had animals at home for the sake of having ample milk, butter, ghee, buttermilk and whatnot for children'* (Fi/22-Durgpura, 2022). Villagers (men and women) repeatedly mentioned reasons for selling animals as open spaces were no longer there, and fodder, although ample on agricultural lands to feed the animals, now had to be bought from the market which was too difficult for them.

The disappearance of good quality, nutritious food from their mats was one of the most important mentions in interviews. It makes sense as food was not just to satiate hunger, but it was the building block of strong healthy body indispensable for work. Repeated sittings with women from different generations at their homes in 2021 were awakening, *'When we cooked a simple chickpea dish, it tasted delicious, now even you cook meat, it is tasteless. Now people use sprays (pesticides/herbicides/insecticides etc), sprays which are nothing but poison, now every crop is dependent on sprays, how can they taste good.'* Further she said, *'No one ever had to worry about food, meat, eggs, milk etc everything was in bulk, people did not have to see a doctor'*. Women repeatedly talked about chemicals as culpable for bad quality food which were responsible for diseases. While talking about the quality of meat, vegetables and grains, they were able to recall not only taste but also the fragrance of rice grains and other cereals. On another instant, one woman said, *'Now food is not the same as it was in the past because now farmers have to put in lots of chemicals and pesticides, in our times the use of chemicals was almost minimal so food was pure'* (FW/1, FW/2-Chachuwali, 2021).

'We don't like to have flat bread (roti) of the flour which is available in markets, it feels like rubber so we use our own wheat flour' (Fi, FW-Kohriyan, 2016). Such differences can only be pinpointed by those who are conversant with what natural food tasted like and the wonders it can do for human health. A farmer in *Burj* (2022) said that waste was an urban problem because in nature everything becomes an input, such as animal waste used as manure tends to produce way better quality yield than the chemical fertiliser. One of the countrymen was telling me a difference in food quality and how the factory processes were turning healthy food into shallow edibles that could satisfy the stomach but were unable to provide the nutrition that the body demands. He used commonly available wheat flour as an example to explain his point. According to him, wheat flour that people used to eat was pure and comprised a major mainstay of our health. When this wheat is processed in mills, they remove all nutrition ingredients (Bhangali, 2021).

Not only the food has been degraded but also the ground water in every peri-urban village has been contaminated and rendered unfit for human consumption since DHA and other gated communities appeared where *gurkiyan (septic tanks)* are still not common; especially in villages sitting on military lands, water quality is still better; nevertheless, the peri-urban settlements have reported every time about degraded quality of water. *'Can anyone believe that village does not have clean drinking water?! This unimaginable has happened to us in almost every village you go to. Development has destroyed everything'* (Chachuwali, 2021). Ground water contamination is now the discourse.

In the past, they might not have pocket full of cash but were inundated with plethora of tangible and intangible assets, clean water and quality food which no amount of money can buy today. Since urbanisation took over their lands for construction of elite housing scheme, it had wreaked havoc on the ecology of the area. The polluted water in every village under the dirt and dust of development-based encroachment tells the story of its failure. '*We had best drinking water just like mineral water. But now water level of clean water has dropped to 450-500 feet (137-172 m), one has to drill a borehole upto this depth to have good quality water and majority of us can't afford the expenditure of drilling so deep. Some wealthy neighbours have a borehole at this depth, though'*, a teacher explained about the prevalent water condition which was not thinkable for rural denizens themselves. *'Now, access to clean water has become a task which has an exorbitant price. One has to have a motor which currently costs 13.500 PKR (44 €) and requires very expensive electricity (people are afraid to receive that electricity bills) then additional PKR 20–25,000 for borehole (82 €-price increases with depth of a bore)* (Fi/22-Dugrpura, 2022).

It is not just the expenditure but an added job to men's daily chores and transport. Men get water either from the nearby filtration plants (see Fig. 5.34) such as *Shareef* hospital, from within a gated community such as Paragon, or from a village where the water quality is still tolerable such as *Mian Singh*, who can afford to get refilled bottles that cost around PKR 50–70 (0.23 €) (Fi/22-Durgpura, 2022). Others acquire water from within DHA or the Nestle filtration plants (Bhangali, 2022). Communities sometime contribute to have a filtration plant to put up in a village such as in

Fig. 5.34 Poor man's burden: People carrying water containers on their shoulders, 2022

Fig. 5.35 Once-multi-purpose small canal passing through Durgpura, 2022

Kohriyan (Kohriyan, 2021). A few filtration plants have been installed by philanthropists where rural people can't afford to have a bore and get water from outside the village, roughly 2–2.5 km/4–6.3 km^2 away (Chachuwali, 2021).

The issue is that a priceless essential ingredient has been transformed into a commodity by modern agriculture and urbanisation augmenting need for cash. The problem here is not just taking a priceless resource for granted, but the environmental destruction (see Fig. 5.35) is a way of making water a commodity for those who had abundance of it in the past. The privatisation of such basic amenities has not only augmented the need for cash but also divided people on the basis of 'who' can have a deeper borehole to be labelled 'wealthy'. In a number of accounts, people found septic tanks as the main reason for the groundwater contamination; *'Water has become contaminated since people started building septic tanks in their houses. The waste water from these tanks is contaminating ground water'*, an elderly woman (FW/1, FW/2-Chachuwali, 2021) recounted the common discourse. The gated communities such as Imperial Garden and Green City built around villages have badly affected the groundwater due to septic tanks in houses. *'The rice produced from this land, when it was being cooked, its fragrance could be felt in the next seven to eight houses, such good quality rice. And as far as water quality is concerned people used to compare it with cow's milk because this area had a very good water quality. Even a person beyond the age a hundred years, would not have a hunch back. Now water quality has deteriorated a lot'* (Poet, FM-Kalasmari, 2018, 2022).

Besides food and water, the open spaces, communal land and spacious houses all had a role in people' well-being. However, the dilemma is that people have lost the agricultural land and open spaces, but size of the houses is also trimming down. People who were the owners of big spacious houses ended up cramping in small spaces as rental income is new trend and people have been converting their houses to quarters for renting out to others (reported by people at several instances (2016–2022)). Villagers in border villages or military lands still have large spacious places, and women find peace on open roof of big houses, lighter breeze and quiet environment (Sarja Marja, 2021). *'We can't imagine moving to the city and start living in small quarters, where one could hardly breathe, there is no electricity right now, but we are sitting on bedstead under tree, and we not feeling hot like one would feel suffocated in cities, this is the beauty of a village'* (FW-Rakh Padri, 2016). The villagers mentioned their relatively large homes in the villages where they could entertain guests, add rooms for expanding family, keep animals and run microproduction units. As one exclaimed *'We have come to (live) in marlas instead of kanals*[8]*'*. The older females were particularly wistful recalling the open spaces where they could keep chickens, plant trees and sit in the sun and shade. They complained that the lack of space for large family gatherings had changed the nature of ties, as now the weddings or other family affairs are being conducted in commercial halls.

The place was contributing to the health and fitness of the people. It has affected women way more than men as they are now totally restricted to houses, getting obese and carrier of several health problems such as knee and joint pain. DHA not only restricted their mobility by erecting walls around them but has also attacked

[8] Local measure of land 30 sq. meters and 500 sq. meters respectively.

Fig. 5.36 Body shifts, soul struggles—keeping past alive: Modern house in Paragon (left), Traditional cot (right), 2022

the landscape of well-being. '*People are getting sick now by being confined. In old times, we used to walk around (in fields), used to sit under the trees which kept us fit and healthy. Now, fans are making us sick. The breeze in the open fields was so invigorating, it is unbeatable. In the past, people put their light bedsteads in the fields where (group of) men and women used to sit in the open, now we (women) are confined to homes only'*, an old woman reminiscence (FW-Dera Chahal, 2021). The families who shifted from a village to gated housing such as Paragon and *Guldasht* had modern houses (see Fig. 5.36) and cars to commute; yet, new environment took a toll, as a woman from *Guldasht* (2016) stated that she got diabetes and hypertension within 2 years of being at the new place, and she still preferred to sleep on the roof at night, *'I feel so suffocated here, there is no life as I had in a village, it's all jammed here. There is nowhere to go, in old times, I could go out with a child in open green fields any time I wanted, now I am stuck here. We (village friends) used to sit at one place, used to eat together, used to go out to buy shoes or clothes etc., there was no dependency on our men. It was so different in village'* (Fi-Guldasht Town, 2016). Similarly, a family who has been living in Paragon after moving out from *Kalasmari* for the past 3 years now says, *'It was easy to stroll and roam around in a village, here, it is only this place (a house), you walk from one cot to another and that is it! We are used to have 'Charpai', we brought it here also with us. Now, I really feel as I am forgetting to walk'* (FW-Paragon, 2022).

In a village, their mental and emotional heath was constantly guided by spending time together. All these added years to their lives. The absence of the Covid-19, especially any fatalities, was a clear manifestation of the fact that rural living protected from urban diseases. A big landholder from *Chak Bhart*, a successful agriculturist who runs his farms and grows food as his business, honestly accepted, *'If you see what chemicals need to be added to increase yield, you will stop eating vegetables, I myself produce cucumber but I don't let my family to eat, it is best to eat cereals only'* (Chak Bhart, 2022). It is not just the quality that makes people nostalgic but abundance as well. The ample organic food which rural people had independent of food markets has now been taken away from them totally. A good food that is the basic right of every individual regardless of social class has now become a luxury. As Christian women (2016) accurately said, *'These malls are only for the rich, these places can do no good for us, we can't even go in there'* (FW-Durgpura, 2016). It is no longer about the healthy food but the provision of food three times a day for (joint) families due to cash nexus, and price inflation is a lot to put up with; as an old woman said, '*Rice, mustard, jaggery, wheat, chickpeas, fodder, every crop was there. Cotton, vegetables, we had everything. And those who did not have, they worked on someone else's land and got what they needed in return, in kind. Now we have bills to pay, several mouths to feed, don't know how to deal with this. Should we bring food from the market or buy milk, how should we prioritize'*. Further, she shared her concern that how knackering the task it had it been lately to place food on the table, and most of the basic food items which were integral part of their diet had already gone including *lassi* (buttermilk, high-nutritious summer drink, made up of milk, yogurt, sugar and butter): '*Old times were way better. Animals were there, we cut fodder for them from the fields. Now 1 kg milk without adulteration cost PKR 170/kg (0.56 €/kg) which is nothing but water. Now wheat is PKR2000/mann*[9] *(6.63 €/mann), previously we used to have it from our fields'* (FW-Chachuwali, 2021). Another woman in *Durgpura* (2022) recalled, '*In former years, milk, rice, wheat, were from home, vegetables too used to come from fields. Cash was less but so were the expenses. Love is dying, now without mobile phones, it has become difficult for people to stay in touch'*. It is no surprise that women locate a deep connection between the abundance of food and the culture of sharing and care that defined village life, where open spaces and welcoming homes once served as anchors of harmony and affection.

5.5.2 Urbanisation and Social Class

A graduate student of development studies commenting on the sale of top soil (2016) praised the policy and said, *'The lands of the villages was worthless in money terms till DHA arrived laid out the plots and started buying and transporting the topsoil to raise the level under the roads to make good quality roads for cars and build a developed housing scheme'*. Some other people from villages talked about the 'value' of land and gave credit to DHA for turning land into gold. An established property dealer well connected with the local influential lot considered land as nothing that had now become of considerable worth since DHA set up a market, '*DHA has aggrandized the worth of land, everyone is doing business over this, the entire market is now in realty. We also have come into this and it has brought money to us*

[9]The current (2024) rate of wheat per mann in the open market is Rs. 4000 (13.26€) in Punjab, Pakistan.

and money brings power. Money helps in establishing contacts but relationships die' (FM-Haer, 2016).

The dynamics and perception about land and its value have been fluctuating. Back in 2014, people showed resentment towards the process and pointed out environmental benefits of trees, plantation and farmland. They also vented out indignation towards the State and its beneficiaries. It was also notified by many that people with small landholdings had nothing much to do with land as it couldn't sustain them; therefore, such owners were more inclined to sell (Sikh-Dera Chahal, 2014). Gradually, importance of land increased even for the landless and that too in terms of mainstream economic financial gains. As it was frequently heard, *'People who had even small-sized lands, they sold and bought two or three times more of a land elsewhere. Hence those who had agricultural land, turned out to be the lucky ones'* (Villagers, 2014–2022). By the year 2022, landless people who only had homes were waiting to be asked to sell, *'Look at the house, all the land around this house has already been sold, DHA has placed their poles as a symbol that land has been taken. This person has 3 marla (76 m^2) house, lives with 5 children. Once he gets paid, he can easily buy 10marla (253 m^2) house elsewhere'* (Chak Dhira, 2022). The property dealer also added that now majority of people were moving to far-flung areas where they could buy agricultural land and could keep a residence too.

Established businessman from *Harpalke* expressed hike in land prices (2018): *'The land transaction (in Harpalke) started in 2004, at that time 8 kanals (0.405 ha) were sold for 35–36 lakhs (11550–11,884€) and now prices have gone to PKR 7–8 crore (231073–264,084 €). My own 4 kanal (0.202 ha) land was sold in 8 crore (264,084 €). However,* villages *are still populated, people go to cities only for business purposes. When the land was sold, 8 kanal (0.405 ha) was sold for PKR 4 crore (132,036 €), after selling land they bought 10 Killa*[10] *(405 ha) in far flung areas, some bought plots, some established markets the majority of people spent money in buying luxury cars'* (FM-Harpalke, 2018).

At another instant, a group of property dealers spoke about increased monetary value of land and its effect on blood relationships, *'In past times, the land did not have that high value in monetary terms as it has today, previously, sisters used to relinquish their share but now they are claiming their legal rights in property. Now people don't rely on persons relationship these days but on money'* (FM-*Patwarkhana*, 2016). In an unvarnished conversation with a revenue record holder (Patwari) of *Mauza Liddharh* and a well-established DHA's broker from one of the peri-urban villages, *Chachuwali,* who still had his residency in a village, he talked about irresistible prices but also kept mentioning the aftermaths or cost people were paying, *'Before the arrival of DHA, every person used to have 12.5 acres (5 ha) of land on average, no one had in 'marlas' (a marla is one twentieth of a kanal) but now they are living in 'marlas'. When I started working as a realtor 20–25 years ago, I used to sell 2 lacs/ acre (660 €/0.405 ha) now in 2021 it is 16 crore/acre (528,415 €/0.405 ha).* He continued, *'There is only one fight and that is because of hunger,*

[10] Killa and acre are the same. 1 acre/Killa has 8 kanals, i.e. 0.405 ha.

since people start getting money, they have become busy in their businesses, now no one fights anyone. However, this is undeniable that when people had less money everything was extremely well managed, food was home-grown, dairy too was coming directly from home, people had bigger hearts; now the more cash people are receiving, more greedy they have become. Now everyone misses out on peace but former times were very peaceful because people were more contented with whatever was available to them, now people constantly struggle to multiply the money. This new development has taken time away from people. As relationships develop only when they are given time, this is why bonding has weakened, money has engulfed a lot, people have lost their health to persistent stress and depression, now sugar (diabetes) and blood pressure is extremely common' (FM-Revenue Office, 2021).

One thing was evident that terminologies have their own meanings and perceptions in different environment *'khush-haali'* (happiness), *'maara'* (weak) and *'tagra'* (strong) in rural context, and now with the arrival of newly constructed environment, these nomenclatures carry new definitions, such as *'khush-haali'* is linked up with 'having money', *'maara'* is the person who is at lower rung of social ladder and *'tagra'* is the one who is financially sound and secure. A big landlord drew my attention towards 'change' being brought by DHA in people's lives through inundation of money, *'Every person is dependent on money. People who had small landholdings such as 2 Killa (0.809 ha), they bought 40 Killa elsewhere, people got the amount they had asked for. But, DHA has broken the relationships, now a man thinks that now that I have money I can forge new relationships, sisters who did not get share in property have disowned their brothers. Paternal uncles (two brothers) fought with each other over money. Now, the only thing endeared to the people is money'* (Natha Singh, 2021). A property dealer in the Paragon office (2015) defended the land commodification and projected it as life-turning episode for a layman, *'People happily and willingly sell their lands because the compensation amount which they receive couldn't be earned in their whole life'*. There is no doubt that exchange of land for one-time offer is indeed life-changing decision but for better or for worse? The 'willingness', 'free-will', 'choice' and all such words sugarcoat the exploitative expropriation. For instance, *Chachuwali* is pocketed within DHA sector and its extension. As people reported that as DHA provides electricity, it also knows the tactics to trouble the poor villagers. They need cash to tackle the problem of malfunctioned appliances due to unavailability of proper power and the foul smell due to blockage of sewerage. *'I myself went upto DHA office and asked them to buy my house so that I could leave, but they refused outright by saying that we do not need to buy yet'* (FW-Chachuwali, 2021).

Majority of the people in these villages were either small holders or landless; however, *'In current times all have become landless'* (FM-Mandianwala, 2016). Rural dwellers, who especially were tied to land because of agricultural mode of production and livestock keeping, were more sensitive towards the process of urbanisation and land transaction. A disgruntled old man (2014) once showed his remorse in words, *'This government is only for the rich, who cares for the destitute'*. Similar sentiment was again being heard by a man who ran a small shop on *Barki* Road when the Covid-19 grappled people with fear. But the corona exhumed all buried

Fig. 5.37 Power play and class dynamic: Big Muslim house locked due to dispute with DHA (top); Christian locality at the back (bottom), Jhallaran, 2022

skeletons of capitalism, *'There is no corona in villages but otherwise has made lives of the poor extremely difficult'* (SO-Durgpura, 2021). It is incorrect to exclude social class or hierarchy from the process of urbanisation to answer 'how' and 'who' is getting served. Development is not one-size-fits-all model, it is meant to create disparity and inequality to keep the ball rolling. Therefore, experiences and standpoints towards the process by those who have been caught in this can't be generalised. The land transaction for the compensation money was neither a linear nor a lenient process as it was experienced by big landholders and well-established realtors. *Jhallaran* (see Fig. 5.37) is the village which has become a disputed territory as people are not on board with DHA, and it also comprises of Muslim (bigger houses, better off) and Christian community (at the lowest in hierarchal order). The economic and social background is source of 'voice' too; one of the big landlords who already sold his land and now was running a shop and a hotel at *Jhallaran* was vocal and fearless about everything going on in there, but members of Christian were scared, hesitant and unwilling to talk. The social class, people belong to, was screaming through walls (researcher's personal encounters, 2022).

Mandianwala village had a community where equality in poverty put them together like a rosary. There were no migrants, and neither of them left the village, but agricultural lands were sold to Paragon (another gated community) and to DHA lately. Dwellers may have got desirable amount of money, but real struggle began afterwards. They went through the hoop in form of betrayal, frauds, migration, rejection from locals, use of force to take over land without payment and so forth. *'Madam, this is no progress but our bad luck. We have brought this havoc to our own selves'*. They talked about incidents which hit them as epiphany, *'A man who was a resident of Shahdra but he bought land in our village, afterwards he sold the land, purchased it elsewhere, but when he went there, locals did not accept him. He was crestfallen, his sugar level dropped, eventually he died the same night. Another member from our village, he had two shops at Jullo, after selling them, he bought a land in Laiyyah, and then we heard that he broken his arm and was walking around with a sling around his neck. Ultimately, he left everything behind and came back. All officers, institutes are allies, they pay deaf ear to us. Only those people had success stories in land, who bought it in villages where they had acquaintances or familial ties'* (FM-Mandianwala, 2016). Another encounter validated the relation between social status of actors and urbanisation and its associated gains and losses, *'We (the poor) pay the price for the rich people's comfort, they (DHA) build roads over people's land without even informing, we struggle to fulfill basic food needs, who would confront them'* (FM-Durgpura, 2021). All the agricultural parcels in *Mandianwala* have been sold to realtor for the development of a gated community. It was a village where entire community was tied to agrarian mode of production, and since the urbanisation encroached, people have had to find jobs elsewhere, usually as gardeners, gatekeepers and wage labourers, and rental income gained more attention, and other joined the army of property dealers—a booming business on the periphery. Denizens did not leave their houses; therefore, settlement was intact.

Land transaction started in *Mandianwala* in 2003, and by 2016, all arable land was sold. This phenomenon was not only new, but a shift from moral to cash economy happened in an unimaginable pace. The former life was still fresh in their minds, and rural actors were constantly weighing and questioning what this new development had given them and what they had lost in the process. *'A man who sold his land is tantamount to selling one's own mother. When we had land, you may sow wheat grains, sugarcane, chillies, vegetable, whatsoever, people used to pluck freshly grown crops the fields to eat and it was so satisfying. One can have a mother in life only for once. Hence, when land is sold, we can never have our lives back which we used to have, it is almost like, our mother has died. Our village was such a place for girls, they did not need to be accompanied by men to roam around and no one ever dared misbehave with women, because people used to know each other and had respect for each other in eyes. In such a short span, blood pressure has become common. If people had small parcel of land, they could get more on contract elsewhere, not just land but households had cows to have milk and other dairy products for free. Now we have to buy everything from the market. Land, was not just a dead, immobile piece of earth but was source of respect; we have lost it'* (FM-Mandianwala, 2016).

Despite individual resistance (according to the economic status of the seller) that comes with money, people appreciated the arrival of DHA in different ways. People did have mentioned that selling land provided them money, and who knew how to use the money well became prosperous. *Durgpura* is located on the main road. Due to a huge geographical advantage to the locality, value of land has risen after DHA showed up, and the building of road networks also helped community members in commuting to seek work. The other advantage people talked about was the annulment of *Chaudhary* culture which has been a distinctive characteristic of rural areas. *'Now everyone has become Chaudhary, no one has to serve big landlords, people are doing jobs and making money independent of Chaudhries. The times when one was dependent and worked as a servant, the Chaudhary could call you whenever he wanted, those days are gone'* (Saduwala, 2021; Gohawa, 2021; Fi/R-Durgpura, 2022). Most of the villages either did not have big landlords or they are very few in number, and those landlords who sold their lands shifted their residence too; hence, servants from the village were no longer required. This is where people give credit to DHA for appending a historically prevailing culture. Although most communities expressed their reservations towards *Chaudharies* and lives being tied to one rich powerful household, *Kalasmari* populace had a different viewpoint, *'Chaudhary*

Fig. 5.38 Face of the progress only for the 'chosen ones', 2022

was like a shaded tree, he used to keep you safe from court, police or bullies in any form. If he could not go himself but just sent a piece of cloth to the relevant bureaucrat the message was delivered, that was his power. Losing Chaudhary was nothing less than becoming an orphan' (FM-Kalasmari, 2022).

Figure 5.38 shows the biased character of the development where its fruits and benefits are not evenly distributed and highly class-centric.

For the longest time, it was a belief that DHA only bought agricultural lands and 'never' displaced the settlements. It is true for those villages which are big in size and have a big community such as *Liddarh, Hudiara* and *Bhangali*. DHA erects walls around them, but displacing them is not at all favourable from business point of view. However, it is inaccurate that village settlements did not face displacement at the hands of DHA. *'DHA does not only buy arable land but houses too. Once they make a map for their phases, they don't keep anything on it which is not included in their maps. This entire village has been sold, there, we used to have a big Bazaar, the whole Chak Dhira has been sold'* (Natha Singh, 2021). The marketisation of land has brought an influx of cash that now drives the erosion of reciprocity, quietly reshaping social relations. What has taken deeper root, however, is the belief that money is the ultimate form of power and that a land transaction is a once-in-a-lifetime opportunity capable of transforming one's fate.

5.5.3 Warmth of Relations to Cold Market World

The peri-urban villages were urbanised villages already as they were connected to two main roads, and people used to go to city for work; therefore, city life was never a conundrum to them. However, their rural world was not dominated by money or cash which is why the biggest currency they had was the community which was built over generations (Accounts collected 2014–2022). Their dependency was never fully tied to the market, which is volatile, unstable, and fundamentally unreliable. Their real strength - their "insurance" in difficult times - came from interdependent relations where help, care, and affection were not commodities but practices of reciprocity. These exchanges, rooted in trust and obligation, could never be purchased through cash transactions. It was this moral economy, rather than the market, that sustained households through hardship and held the social fabric together (Chachuwali, 2021; Chak Dhira, 2022). The real-estate business in the shape of creating gated communities is booming by selling intangibles which they can neither deliver nor ensure. The 'knowing' each other is a part of rural culture, which gives them a safety blanket, *'There was a time when one could hardly see anyone at night and people would feel scared, and now within 7-10 years, there is no difference of day and night. When I set up this small shop here, a woman came up to me and congratulated me. I was unsure of her reaction and asked her a reason, to which she responded, 'Brother, now if we come to this road at 10 at night, we would know that a member from our community is sitting there already'* (SO-Durgpura, 2021). I personally have experienced that how beady-eyed community members could be and how readily their brains could spot an outsider from afar, no matter how meticulously he tries to blend in.

Now, their world is changing, and they are trapped in the process of development where they can see where the true 'value' lies. Interlocutors were from varied social classes, apart from rich landholders, and denizens who gained advantage from land transaction shared the meaning of quality of life which was not different from the ones who were at lower rung of social ladder. Despite material gains, villagers unanimously lamented over the loss of intangible assets which used to determine their quality of life, and no amount of money could ever buy those assets back, '*It was a stress-free environment back then, everyone used to stay connected. Blood pressure is not a disease, it is a result of disturbing stream of thoughts (to acquire more and more). People had a good life in old days, used to spend time together, if I had cooked something, everyone could join and take a bite, women making breads in clay-oven, everyone is eating together, buttermilk was in bulk. We have lost mental peace now. This pretentious world is nothing but a lie, fraud. Community was fraught with love and empathy now relationships are open for bidding*' (Fi-Chachuwali, 2021). There were resented reports by people throughout fieldwork at multiple peri-urban villages about the loss of healthy dietary intake, water and air got contaminated, people were forced to sell their animals (source of dairy and organic food) due to non-availability of the communal land and agricultural turfs, and increasing sense of insecurity again due to fragmentation of the community (FW/1-Chachuwali, 2021; FW-Paragon, 2022; FM-Kalasmari, 2022).

Women had a huge contribution in order to provide understanding about how the process of development through urbanisation unravelled itself. There is an on-going debate revolving around women's safety and harassment and the exercise of power over women by men (patriarchy), and talking to women in rural settings elicited this different set of experiences. It revealed that harassment is an urban issue, and restriction from men is their concern for their women's security instead of oppression, '*We are bonded to each other in harmony and love, be it marriage or be it any difficulty we all stand together. Another thing, it is a very safe place as everyone knows everyone. We women never feel afraid of being harassed if we go out alone*' (FW-Dahoorwala, 2021). The discussion with women in border and peri-urban villages and enclosed settlement unties the role of a woman herself as a carrier of an ecological set of relationship not only within the family but also within the community. Women also act as a rosary that ties family and community along with livestock and land (FW-Heera Singh, 2016; Burj, 2022). Spending time with them burst many myths that I had carried whole my life in a metropolitan city such as they were not found under household chores duress, household duties were reduced as compared to before and animal keeping was also not totally a woman's responsibility. In fact, women and men used to take care together, or men were the ones who handled the hard part of this task (Women, 2021; 2022). Young girls are studying with no pressure for taking over household duties, which is one of the advantages of joint families.

The women in peri-urban villages shed light on several social issues such as inflation, adulteration of food and water and how the loss of open spaces has affected their health badly, and they raised the issue of greed with the arrival of DHA as it brought people opportunities to have financial gains, distancing people by engendering inequality and its corollary effect on finding suitors for children' marriages. Above all, they spoke of a sense of security that village life had long provided—a security so deeply rooted that women never felt the need for male accompaniment, and where marriages were lived as partnerships rather than as constant negotiations of authority. In the village, they saw themselves and their spouses as part of a shared team, not two individuals locked in a quiet struggle for dominance. This everyday ease, forged through familiarity, mutual trust, and a collective moral world, stands in stark contrast to the anxieties they now face in the transformed landscape around them, '*People are getting sick only by being at home. In old times, we used to walk around (in fields), used to sit under the trees which kept us fit and healthy. Now, fans are making us sick. The breeze in the open fields was so invigorating, it is unbeatable. In the past, people put their light bedsteads in the fields where (group of) men and women used to sit in the open, now we (women) are confined to homes only*' (FW-Dera Chahal, 2021). Therefore, for these women, losing the community is tantamount to losing roof over one's head. The relations they built over time by sharing space and time gave them sense of belonging and identity (Kalasmari, 2022), which holds them back from moving to better-off localities, '*A major question arises in mind is that where to go after selling the house and what if we wouldn't get the environment that we have here*' (Fi/22-Durgpura, 2022). However, people were also very open about the cracks already appearing in relations ever since money had taken over lives, and even blood relations were drifting apart. '*There is no warmth or respect in relations anymore, people only worship money*', this was heard time and time again throughout.

5.5.4 Conclusion

The direct conversations with people over the years and personal encounters at places revealed the impact of urbanisation, which is projected as 'development', is 'hollow', but its damages are not curable. The urbanisation on the periphery where villages as landscape and villagers as marginalised communities experiencing it has unfolded the true character of capitalist development. This chapter has shared the narratives and experiences they have had with 'development' from wide range of actors, coming from villages which are at different stages of development simultaneously. Hence, there is a gamut of experiences being encountered on material, social and psychological landscapes. Moving across different classes, genders, generations, and ethnicities through multi-sited ethnography was a deeply felt endeavour. It allowed me to grasp the subtle textures of everyday life and the layered nuances of change. Each site functioned as a microcosm of wider socio-ecological transformation, revealing how neoliberal development projects take shape on the ground - unmasking their logics, their fractures, and the quiet ways people endure, resist, and remake their worlds.The policy makers, planners, developers, city officials and realtors measure development solely in economic terms and physical landscape to replicate the modern western world. This chapter sheds light on the quality of life and its essence for rural people which they repeatedly mentioned that what they have lost in the process can never be compensated by any amount of money. The research brought those on the forefront who are invisible as they are not deemed to be significant in the process, and their money-poor status renders them valueless on so many levels that are not considered to be seen. It also accentuates the fact that reality of development as a project and urbanisation as economic growth policy in human and physical monocultures can only be understood through the lens of those who are paying the cost of it.

Village peeps not only found 'money' culpable for the loss of reverence and other intangibles but also found themselves trapped in a vicious cycle of money-making that made exchange of land enticing, '*Villagers are themselves waiting for DHA to take over their land, because they used to do farming and now that 'mahoal'(environment) has changed, so gone are their aspiration about agriculture now, this area is no longer the same*' (Fi-Chachuwali, 2021). People in the poor enclaves are still holding on to what is left, '*We still have a good environment here, we all know and meet one another with warmth. This is a fact although that money can't buy the happiness but it has now become a necessity too*' (Fi/22-Durgpura, 2022). The subtle violence of cash has effectively battered the defences of the poor and cash-hungry farmers and in the process took away the base of their social cohesion, land. Through demands and needs for cash have been rising—for modern education, medical emergencies, electricity, fuel, debt repayment, dowry, children's marriages and much more—even this newly acquired wealth may not last. What was imagined as a life-changing windfall already shows signs of thinning. The very money that promised security now feels fragile, unable to sustain a future shaped by escalating costs and growing uncertainties. Beneath it lies a quiet grief:

Fig. 5.39 Dear rural ecology, rest in peace… IT IS OVER!. Ecology's burial ground, 2022

communities who once lived without fear of scarcity now confront a precarity that money cannot mend, a slow unravelling that is as painful as it is invisible. Not only through reports by personal interaction with the ones who got great deals and became millionaires overnight but also through common villagers joining hands as relators, despite a few success stories, they shared the similar narratives of loss and gain account when it came to intangible and tangible assets, *'Development has come with a cost, food, water, health, relations, courtesy – all have got ruined'* (FW-Bhangali, 2022). It was found that they had a clear line between wealth and well-being, being rich and happy or contented (Villagers 2016–222). Fig. 5.39 should be seen as the representation of environmental destruction, dismantling of rural realm and converting relatively stable zones into destabilised zones for producing monocultures. This picture also delivers the chaos and damage done to the world that can only be felt, as certain landscapes, e.g. psychological, emotional and cognitive landscapes, cannot be seen, but one can solely try to understand by being in contact with those who are facing it.

References

Javaid, H. (2023). Can money compensate? Gain and loss accounts from urbanizing villages of south-eastern Lahore, Pakistan. Heidelberg Asian Studies Publishing. https://doi.org/10.11588/hasp.1166

Javaid H (2025) Can Money Compensate? Land commodification, urbanisation and social change in peri-urban villages, Lahore [Heidelberg University Library]. https://archiv.ub.uni-heidelberg.de/volltextserver/id/eprint/36529

Chapter 6
Giving Away More than Land: Neoliberal Urbanisation and the Erosion of Community Life

Contents

Abstract This chapter examines how neoliberal urbanisation in peri-urban Lahore dismantles not only agrarian ecologies but also the social fabric of community life. While land acquisition is justified through compensation packages, these payments conceal deeper impoverishments: the stripping of fertile soils, loss of commons, fragmentation of kinship networks and the erosion of cultural practices that once sustained collective resilience. Ethnographic insights reveal that compensation produces new hierarchies—turning farmers into small-time real-estate dealers, women into precarious domestic workers and children into consumers detached from agrarian ecologies. The analysis situates land not as a passive commodity but as a living ecology and a moral economy, whose dispossession reverberates across generations. Communities that once thrived on reciprocity and subsistence are remade into fractured, cash-dependent households, increasingly alienated from one another. By tracing these transformations, the chapter argues that neoliberal urbanisation is not merely the conversion of land to concrete, but the systematic unmaking of life-worlds—where compensation is offered, but community is irretrievably lost.

H. Javaid, *Land Commodification, Urbanisation and Social Change in Peri-Urban Villages, Lahore*, Advances in Asian Human-Environmental Research, https://doi.org/10.1007/978-3-032-10462-5_6

6.1 Discussion

The neoliberal economic model has come with a huge cost with regard to environmental damage, but dismantling of social structure has a domino effect on people's lives. The conversion of arable land into urban land has several environmental and social implications. Simon (2008) mentions the issues, challenges and opportunities due to urbanisation in peri-urban villages or regions of rural–urban interference. He stated loss of cultivable land not only coupled with degradation but also has jeopardised local food reliance and production of surplus to be sold to urban dwellers. Land acquisition is fundamental to many forms of accumulation, as mining operations, monoculture industrial farming, infrastructure building, gated communities, etc. are regarded as 'development' with a promise of economic growth. The World Bank, governments and their biased polices play a significant role in twenty-first-century colonisation through large-scale land acquisition (for development of already developed part of society). This in turn causes tangible and intangible impoverishment, loss of commons and quality of life, jeopardising communities (particularly those living on state-owned land and have weak legal position) especially by denying their rights to traditional ways of living (Denis 2018; Hall 2013; Kröger 2012). This is relevant to an extent where people are living on the periphery which is suburbs of the poor, and majority of them are not only from the lower economic background but also have weak or no legal documentation of property. People who are on state-owned land are relatively safer than those who live on privately owned land but are not financially strong.

It was also noticed in the villages of Lahore that people eventually had to give in due to weak economic background and lack of means to put up resistance and to exercise their agency when it came to land transaction. However, the ones sitting on state-owned land were at dual advantage, as they were far from land appropriation, still one of the members of the family living on state-owned land, obliged to continue farming, which provided them two-third of the produce, and people could stock grains, keep the animals and have farmlands to have fodder for animals. A significant proportion of households relied on milk selling but also supplemented their earnings through urban jobs, rental income, or non-farm employment within the family. This diversification strengthened their economic resilience. Consequently, communities situated on government land were comparatively more secure, whereas those residing on privately owned land faced heightened insecurity due to weaker income buffering and greater exposure to land market volatility. However, this also can't be said with utmost surety as *Jindra Kalan* is also on military land, but some part of it has been sold to build DHA gated community and has been walled off from the side that connects DHA residential society. People can vividly recollect what was right in front of their eyes till 2007–2008, and what came afterwards reminded them of loss of assets and a sense of rejection. Similarly, *Chughalpura (Habibabad)* is another example of such village on military farmlands which is a little far from the main road that makes it isolated, and even that land has also been up for transaction but illegally. Village communities are certainly not homogenous, and the heterogeneity is coming not only through their economic backgrounds but also through the sect they belong to and the religion they follow. For instance, Christian community can be found in every rural settlement where people are at the lowest in social hierarchal structure, as they were mostly landless, and being a part of minority somewhat became advantageous for the developers. They are easier targets not only to expel but also to create rift and doubts amongst the community. It is nothing less than killing two birds with one stone.

Goldman (2011) takes Bangalore as a case study to argue the role of state and its policies to facilitate speculation-based activities which cause rural dwellers to displace especially due to insufficient proof of their ownership. The state's aim of creating world-class cities or 'megacity syndrome' (Mpungose and Maharaj 2022) has irreparable effects. This study has taken a village as microcosm of socio-ecological change to study the impact of obsession of making world-class cities in order to open floodgates for foreign direct investments (FDI) and local also. It helps in disentangling capitalism and now neoliberalism as economic models in creating monocultures. States are unwilling to recognise that urbanisation as development model can't be taken as one-size-fits all not only in regards to social structure and communities but also neglecting environmentally stable zones, treating them as disposable unless they will be converted to concrete jungle. This practice has been failing mankind. Qadeer (1996) explicitly talked about different phases of urban development plans in Pakistan and the era of 1970s onwards (commenced with up scaling of Lahore Walled City) as the beginning of inference of international bodies such as the World Bank, through urban policies. The World Bank's funding as 'aid' was nothing but a noose around recipient's neck, where receiver was supposed to abide by policies and models. Further, he underscores the period of the 1980s and 1990s as a shift to private sector for development, dictated and pressurised by the World Bank and IMF. In the 1990s, when cooperative housing societies became major land developers, urban land markets were booming and speculation raised the bar for high returns. The multi-lateral donors like World Bank, big endorser of myopic, profit-centric, selfish game of capitalism in the name of development through infrastructure building in Global South (Steel et al. 2017) were only rendering fringes as a *Tabula Rasa* (Kennedy and Sood 2016). The idea behind making a 'world-class city' is not only an image building but also a form of colonisation that dictates developing countries to prove their worth by building the replicas of developed countries. This is the only parameter to prove that they deserve to be 'controlled' by international financial institutes. Lahore has lost large swathes of agricultural lands and social landscape, losing social capital as a result of land and resource grab. For instance, the cruel scrapping of soil from agricultural lands to dump it elsewhere for road building and high-income enclaves amicable for the rich; letting agricultural lands die once it is enclosed and complete removal of tracts of productive farmlands for the construction purposes further validate the capitalist treatment of land as nothing but a *Tabula Rasa* and also the dismissal of social and ecological landscapes. The transformation for the sake of city expansion has been promoting consumerism at the expanse of environmental degradation and aggrandising social ills such as targeting weaker and making them even more marginalised, displacement by pushing people out of their own land, breaking communities to deplete social capital, making gateway for the emergence of diseases and so on. This research has witnessed that development policies and city planning work in a way that supports the high-income faction of the society which ensures miscarriage of social justice.

Denis (2018) discusses the urban processes on the periphery in GS since the liberalisation phase has invited gamut of actors including the poorest household and working classes who are eager to cash value of land they own. Especially, the financialisation of land that does not involve development of the land itself but can be put up for exchange in market in land makes land desirable and favourable for the real-estate growth. The peri-urban region bears the load of urban expansion the most in recent time, and it is also a tool for land speculation nurturing real estate (Narain 2009), facilitating market in land across the countryside (Banerjee-Guha 2009; Levien 2013). The treatment of land as a commodity has turned the periphery into a landscape of intense transformation, foregrounding what scholars describe as "agrarian urbanism", a framework that highlights the intimate entanglements between agrarian questions and urbanisation (Gururani 2020; Gururani & Dasgupta 2018). In this view, the rural–urban interface is not a zone of linear transition but a space where long-standing agrarian relations, of land, labour, tenure, and livelihood, are reworked through the pressures of capital accumulation and speculative development.

Peripheral urbanisation, as theorised by Caldeira (2017), emerges not as a by-product of metropolitan expansion but as a mode of urbanism in its own right, marked by fragmentation, informality, and uneven governance. In South Asia, this resonates strongly with Ananya Roy's (2009) formulations of "urban informality" and "new geographies of dispossession", where state power and speculative capital intersect to recast agrarian landscapes as frontiers of accumulation. Pakistani scholarship has documented similar dynamics. such as Matthew Hull have shown how land informality, bureaucratic arbitrariness, and the state's discretionary power shape property regimes. Akhtar and Rashid (2025) and Gazdar (2009) highlight how land markets, patronage networks, and tenancy relations create layered vulnerabilities in peri-urban Punjab. These regional perspectives underscore how agrarian and urban transformations in Pakistan operate through hybrid political economies involving village elites, local bureaucracies, military-run real estate institutions, and private developers.

These dynamics generate forms of agrarian dispossession that political economists have long identified (Harvey, 2003; Levien, 2018) but in new and hybridised ways. On one hand, farmers are converted into small real-estate intermediaries or property dealers, navigating volatile land markets and speculative cycles. On the other hand, these processes create what Chuang (2015) describes as an expanding army of "proletarianised workers", people who lose access to land-based production and are pushed into precarious, low-wage urban labour markets. This mirrors findings from South Asia, where scholars such as Sivaramakrishnan (1995), and Rita Brara (2006) have documented the rise of rural-to-urban precarity as agrarian communities become absorbed into informal labour circuits.

This dual transformation resonates with Bernstein's (2010) formulation of the agrarian question, in which commodification restructures class relations and blurs distinctions between rural producers and urban labourers. In peri-urban Lahore, this manifests as the dismantling of agrarian livelihoods and the emergence of new class positions, speculators, brokers, wage labourers, and informal service workers, layered onto older caste, kinship, and zamindar–kammi hierarchies. Pakistani anthropologists and sociologists such as Saba Mahmood, Irfan Ahmad, and Alavi (1973) have similarly illustrated how class, kinship, and religious identity mediate everyday governance and shape responses to land transformation.

The political ecology of this transition reveals how state policies, planning regimes, and capital flows produce differentiated vulnerabilities (Swyngedouw & Heynen 2003). In Pakistan, scholars such as Matthew Hull (2012), Ammara Maqsood (2017), and Danish Khan (2022) have shown how uneven infrastructural development, speculative housing societies, and spatial commodification of land through schemes intensify alienation, exclusion, classed and gendered inequalities. The commodification of land not only redirects ecological resources toward elite consumption but restructures social worlds by replacing long-standing moral economies with market-mediated dependencies (Scott, 1977; Thompson,1971), a process mirrored in South Asian critiques of neoliberalism (Baviskar, 2003).

Seen through this expanded theoretical lens, peri-urban transformation in Lahore is not a neutral evolution. It is a deeply political process shaped by state–capital alliances, differentiated forms of dispossession, and the reworking of agrarian identities and class relations. It is in these contested spaces that the stakes of Pakistan's urban futures, and their attendant inequalities, are being negotiated. People who were into agriculture and milk selling, especially in pocketed villages, are serving as service sectors of the elite with jobs which have no guarantee such as car washing at people's residence as per demand, or as a domestic help has become very common amongst women, and even villagers themselves have started hiring other women to work as maids in home which was not a common phenomenon in a village otherwise. The big landlords used to have staff, but now even small landholders or people with some form of better source of income prefer to find maid that is becoming easier due to large pool of women as house help to pick from. This also allows household in gated communities to be exploitative as they have a queue of women who are ready to work at lowest possible rate of household chores (especially the ones who have migrated from other areas). This reminds me of the fact that women's rights and their ill treatment can't be understood and resolved unless we look at it from a class question. Capitalism has been thriving on dividing societies into distinctive groups based on who haves and have-nots; therefore, feminism can't be put separate from the process that has persuaded women to be available to sell their care and affection for corporate working women on salaries (Comanne 2021; Malatesta 2019). The general perception regarding women in villages as suppressed and being trapped in patriarchal clenches also came out different. Their marital relationship was a partnership to them in lieu of power relation.

The new development process, urbanisation to be precise that has been happening at the loss of agricultural land and transforming it into elite gated housing societies, is seemingly a linear process, but it has a lot to it. It is a shift from an agrarian mode of production (MOP) to the neoliberal. A former MOP was the reason of subsistence absence of which has now made people totally dependent on market economy which is highly volatile. This subsistence was attached to the structural organisation of a rural ecology that despite its imperfections was giving basic shelter to all. Land as a means of MOP has had profound importance, and this is where Britishers changed the dynamics through the introduction of private landownerships and made half of rural population as landless (Bhattacharya 2019). Grover (1963) stated that the State was the proprietor of jungles and wasteland for agricultural purposes, but it did not exercise its ownership over the cultivable land which was already in the possession of the *Raiya* (peasants). State claimed share of the produce but not the entitlement to the land. Moreover, no serfdom that tied peasant to land existed during Mughal age, and agricultural land was not considered as a saleable commodity the way it became in the late eighteenth and nineteenth centuries. It was an epoch that fortified the unjust land distribution since then, rendering all artisans and landless *ineligible*; hence, it left them out of the race of people demanding the land rights. But it favoured the landowners in elevating their wealth at the cost of those who were made invisible and voiceless. The primary issue lies in the land distribution system which, according to Punjab Land Alienation Act 1901, clearly divided rural population into agricultural and non-agricultural or cultivators and non-cultivators under the colonial rule, and agricultural labourers were reduced to mere wage labourers. This continued during Punjab land reform that decided to allot land only to tenants who did not own their land or cultivators with small landholdings. This resultantly yet again dismissed the class of non-cultivators (Gazdar 2009). The exploitation was further intensified with the arrival of green revolution in the late 1960s, which benefitted the ruling classes, and according to Ahmad (1970), it converted big landowners into capitalist agriculturists. Hussain (1984) also pointed out that how capitalism-led agriculture robbed small farmers of their freedom and made them completely dependent upon big landowners for loans, agricultural inputs, etc. Moreover, the market gave more power to landlords due to their political and social position to act as mediator. The capitalist farming shifted poor peasants from sharecropping to cash rent. Due to which, to combat with increasing cash rents and debts, they had to sell a portion of their subsistence harvest output at low prices. Despite all this, the new development on the fringes opened a gateway to new form of capital accumulation which seemingly appeared to have a chance to better life and to get rid of agriculture with no returns. However, the research put emphasis on the fact, which was voiced out by people time and time again, that cash could be the gain, but the intangible assets they had been losing in the process, including their identity, are gone for good. They also have a strong realisation that they could not be able to adjust in better-off gated communities as neighbourhood would remain unwelcoming for them. This widely observed pattern reflects a deeper spatial logic of displacement. People gravitate toward familiar villages or village-like environments in an effort to reproduce the social worlds that once anchored them. Yet even when such groups establish new settlements, the structural conditions that precipitated their movement remain largely intact. As households relocate and begin inhabiting a new site, their collective practices, kinship routines, labour exchanges, moral obligations, and everyday sociability gradually reconstitute the space into the form of a new village. In this sense, the village is not simply a place but a social formation that is continually recreated, even as the broader forces of dispossession, insecurity, and uneven development continue to shape and constrain it. *Ghawindi Khush-haal Singh* is one such example where Christian community shifted from *Bhangali* and started living in a new settlement.

The use of land acquisition laws, converting private property into 'public purpose' by the State with the addition of never-be-able-to compensatory package, is a way to legitimise the land and resource confiscation with the characteristics enclosure, erasure, eviction, exclusion and extraction from the Global South (Neef et al. 2023). From the vantage point of the city, a parallel dichotomy emerges. It is visible not only in the spatial and social stratification of the urban core, but more decisively in the ways peri-urban landscapes are systematically extracted, remade, and annexed to sustain elite consumption. These extractive practices reproduce, at the micro scale, the very logics that shape relations between the Global South and the Global North, where uneven development, dispossession, and resource flows are normalised as the cost of growth. In this sense, the city's peripheries become internal colonies, sites where global hierarchies are territorialised within the nation's own urban frontier. The land deals in struggling economies have a lot in common with peri-urban villages of Lahore, but the attribute of 'erasure' (Devine and Ojeda 2017; Neef 2021) is striking. The authors are talking and exposing the side of tourism, the creation of unequal power relations between host and guest and interwoven forms of violence in the practice. The purpose of mentioning it here is that once these villages were rich with history, i.e. the stories of partition, the turmoil these places went through, the symbol of resilience against the times and the ability of a place to embrace who wished to reside. They had their identity and recognition due to their cultures and tradition that gave them a sense of belonging. The removal of these places, including those that are sentimental to communities especially graveyards and shrines (used to carry name of a village), is a direct measure to wipe the slate off. It makes me realise that how sorted rural people are when it comes to their priorities. They cannot fight every battle, and when they need to pick one, they do it not on a whim but wisely. It reflects through their stern stance and firm unwavering resistance to guard their graveyards and shrines. Even the earliest times when communities succumbed to unimaginable prices of land transaction, the same community of *Charrar* fought tooth and nail to protect their graveyards. Although people's perception has been transforming and going through restructuring, their emotional attachment to the graveyards—a place for burial, ancestral strings and a source of identity and recognition, still remains intact. Still, there are villages who sold their graveyards such as *Heera Singh, Mota Singh, Sangatpura* and *Jalal abad.* The ethnicity also plays role as it also ties with socioeconomic status. For instance, Christian community is more prone to selling graves as happened in *Bhangali*, where Christians sold a few graves. As they already are at a lowest social rung in the village, landless, therefore, their urge for gaining as much as they can comes out stronger. On the other hand, Christians in *Durgpura* did not let DHA takeover their graveyard. Hence, the use of 'cash-compensation' as a tool works for the realtors and seemingly for sellers in a way that capitalism justifies the land grab and dispossession by projecting it as a life-changing event for the ones who could never have imagined it. For instance, setting up land market and converting farmers into small realtors and to camouflage social injustice via 'compensations' (Alias and Nasir Daud 2006; Ghatak and Ghosh 2011; Mahalingam and Vyas 2011; Mallik and Sen 2017; Sarkar 2014, 2015; Vijayabaskar 2010), to bring ease to violators engendering inegalitarianism (Tassadiq 2022). The study on peri-urban villages also confirms that these 'compensations' are far from reparations, and making up for losses of all kinds, especially the intangibles, is not possible. Also, these compensation packages are nothing less than 'redeem packages' to justify environmental degradation and fragmentation of communities, and taking away of land that was meant for agrarian production and source of subsistence can mean nothing. Also, a land that has a social character too determines social relation within and across hierarchies where dependence on people is the key, and the intangible assets which are a product of historically created ecosystem determine the quality of life for rural/peri-urban communities. On the other hand, the land commodification has put an end to *'Chaudhary'* culture with which most of the village communities are contented to be their own bosses and has created the presence of absentee landlords. The land transaction brought bliss to the ones who knew how to use money wisely, and this made the business in land as a realtor flourish.

The project not only has seen how different rural and urban realms are in multiple ways but also tends to see how and why villages which were considered backward regions are stable and sustaining, and communities are richer in ways which modern economic world can't process. Another character of cash compensation is that realtors find it just as owner is getting an 'offer' only, and he has liberty to decide whether he wants to sell or not. It has persuaded me to think that how tricky this word 'choice' can be. Who can exercise 'choice' in truest way is primarily dependent on persons' socio-economic status, especially, when it is negotiated between two sets of people with extreme power imbalance. Therefore, urbanisation is not merely a process of producing new terrains of uneven geographies; it also ruthlessly dismantles socio-ecological worlds. (Denis 2018; Swyngedouw and Heynen 2003; Vij and Narain 2016) that make an 'environment' and especially the brunt on periphery due to the development through intense rural reconfiguration.

The peripheral urbanisation has also been called land grab/land rush (Neef et al. 2013; Neef and Singer 2015; Zoomers 2011) and gentrification (Davidson and Lees 2010; Farahani 2013; Smith 2010), and many have seen through a framework of accumulation by dispossession (Doshi 2015; Magdoff 2013; Shin 2016). Swyngedouw and Ward (2020) explain the social character of land, which for market is just a commodity, a fungible thing, but as a matter of fact, it is attached with value, symbols, meanings, culture, emotions, social organisation—all such embedded traits which make land not just a laying piece of earth but very much living and central to ordering of life in a community are ruled out from capitalist accumulation process. Also, etymologies play a role in making processes legit, holding different meanings on the basis of 'who' is looking at. In rural areas, land is the fundamental to all human, human–nature, livestock and biodiversity relations embedded in agrarian ecology. Land is static and immobile but has the ability to produce not only food but also relations. Farming itself is not just a livelihood but a way of life, which is why losing arable land is far more than just losing livelihood; it is a loss of life. Development treats geographies as a chess board where moving a piece from one square to another does not affect a piece itself. This is how planners, realtors, policy makers and the state believe and treat communities, which in reality devastates people emotionally, mentally and physically. The removal of land from a village is a tipping point. Safransky's (2017) analysis speaks directly to the dynamics unfolding in peri-urban Lahore. Her work shows how state actors deploy the category of "vacancy" as a political instrument, using it to legitimise redevelopment projects that erase existing social worlds. This tactic parallels the way Pakistani state institutions and developer–military alliances classify agrarian land as "idle," "unproductive," or simply "available for public purpose" to justify its conversion into elite housing or infrastructure schemes. Just as Detroit's farmers and long-term residents contested the notion of their land being "vacant," villagers on Lahore's urban frontier similarly challenge official narratives that deny the histories, labour, ecological practices, and kinship networks embedded in their landscapes. Their insistence that the land is inhabited, cultivated, and socially alive exposes how state-led urbanisation relies on erasure - the deliberate overwriting of lived spaces with technocratic descriptors that serve capital accumulation.

These struggles reveal a shared political logic across very different geographies: whether in the Global North or the Global South, the power to define land is inseparable from the power to dispossess. Both cases illustrate how urbanisation is not merely spatial expansion but a deeply political process that determines whose presence counts, whose claims are recognised, and whose worlds must be dismantled to produce new terrains of profit and enclosure. The study also has seen closely that how words, meanings, perception and value towards tangible and intangible assets but the outlook toward a certain process change based on gender, class, social standing, livelihood, association and the way someone establishes the connect. This also insinuates that the power dynamics, be in relation to money or in the form of land that is tantamount to the power and influence, plays a significant role not just between a community and a realtor but also within the community. Which is why, in urban political ecology, such development is indispensable for capital accumulation, creating a sphere for environmental processes that negatively affect some social groups and benefit others (Swyngedouw and Heynen 2003).

Levien (2015) through his study on special economic zones (SEZ) in Rajasthan argues about the role of individual social capital in making farmers as successful brokers at the expense of fellow villagers. Thesis does recognise the misuse of trust by an individual as a broker or group of individuals who were not the investors but amongst the influential in the community and partook in selling communal land against the wish of the rest and did it without informing the fellow community members. However, it can't underestimate or negate the power of collective social capital which is the only safety blanket and source of resistance against the big guns. One of the losses they lamented the most is the diaspora or a loss of a community. The point to ponder is whether displacement is the only way to break the community? The answer is, 'no'. The inhabitants have started feeling a certain disconnect amongst the community members since money has invaded as one can now vividly see the financial background of a member which has created a distance amongst people. As Marglin (2008, 2010) unequivocally expounds the impact of market economy in dismantling the stability of a community through instilling greed to gather more, individualism and by making inequality very obvious. Communities who had a symbiotic relationship with the places they inhabited on were bound to take care of the commons as they were dependent on them for the livelihood and to meet needs of their day-to-day life. This harmony with nature was coming from the collective and communal consciousness, which developed their sentiments towards the place. Indigenous would give 'value' to a place is utterly a flipside to what economists or developers or policy makers define 'value' as. Not just the livelihoods but emotional attachment to a place give rise to insurgence and revolts and persuade people to fight till the last drop of their blood. When I look at the villages which mostly no longer fit into this box, I observed that the surrounding area has started to transform, despite being the same place, it is causing alienation and detachment, and gradually and slowly people are getting detached which doesn't give them a solid reason to hold on to. Tragedy of the commons can only be suggested by mainstream economists as commons and privatisation are not only the opponents in the same ring but latter is the fundamental to flourishing market economy. Tragedy of commons is not the locals but external stimuli in form of urbanisation. The contamination of village water resources is one such example. The state's paternalistic relationship to land, inherited from colonial regimes and reproduced through postcolonial land settlement, redistribution, and regulatory acts, continues to shape the politics of space and belonging. These legal and administrative instruments, ostensibly designed to manage land acquisition, sale, and transfer, have become mechanisms that entangle communities in cycles of dispossession. In doing so, they undermine the ecological integrity of landscapes and destabilise the social worlds that have long depended on them. The consequences extend far beyond any single development-impacted locality. They spread outward like a fire in dry forest, uncontrollable and indifferent to the lives it engulfs.

In this regime of accumulation, agricultural land is treated as a gold mine. Yet what unfolds is a paradox of extraction without value. Developers strip the earth of its living richness only to bury it beneath asphalt, raising road levels to save negligible sums in fuel and transport. The soil, once fertile, breathing, and shared, is rendered lifeless. What could have sustained generations becomes a carcass of its former vitality. This transformation is not simply environmental degradation. It is the conversion of socio-ecological commons into inert urban surfaces, a testament to how the promises of development are built on the slow death of the landscapes that make life possible.

The development through urbanisation in peri-urban villages has been an eye-opener to make us realise that how far these flawed policies and ruthless economic models have gone to crush the masses of the society to cater advanced stages of capitalism. Abundance of food has become a history, quality food is in dreams, food security has been jeopardised and whatever we have left in the name of food is nothing but a poison. The modern agricultural techniques have turned agriculture into hub for disease-engendering areas and have forgotten that the earth is a living being and needs its own time and space to rejuvenate after the harvest just like a mother requires rest to heal after birthing. The inference of 'modernity' to food in form of fast foods and bread and renouncing traditional way of eating healthy has already been proving that modern and being healthy can't be found in the same package. Yet, access to good quality food now goes hand in hand with one's economic status and social standing. Likewise, water has not only been contaminated, it has become a new financial status-tester for community members. Now, access to clean water, which was freely available and was in abundance, is now in reach of those who can afford it. The enclosure- and exclusion-based neoliberal development that is also an era of neocolonisation is the resurrection of violence and exploitation historical literature is filled with. The ongoing process commits violations at multiple levels. It appropriates land as a resource and deprives people of the capacity to grow food that once sustained large families with ease, turning dispossession into a form of food theft as well. Cow dung, previously used as manure and a natural fertiliser within village ecologies, depended on the availability of land. With land now out of reach, even dung has become a commodified item sold through Amazon and other global e-commerce platforms. The everyday infrastructures that anchored rural life have also been dismantled. Every village once had at least one communal water well, yet these have disappeared, and clean groundwater, once abundant and trusted, has become scarce. In response, men now take time out of their workday to fill containers at filtration plants, an infrastructural burden villagers never needed in the first place. Together, these transformations reveal how dispossession reshapes land, labour, and lifeworlds, undermining the ecological and social systems that once enabled rural communities to sustain themselves with dignity and autonomy.

Meeting women in peri-urban villages introduced me to a flipside of what people normally heard of women in such areas. It can be from strict gender roles to stories of oppression and control by men and minimal-to-none say in decision-making, especially economic matters (Akhtar et al. 2019; Petesch and Badstue 2020). Women have long been recognised as deeply connected to ecological systems and as possessing a nuanced understanding of their environments (Agarwal 1989). Their everyday labour, particularly in tending livestock, gathering fodder, and managing the rhythms of household production, places them at the centre of rural socio-ecological life. This also means they are among the first to experience the consequences of dispossession, especially when land that once sustained these practices is reclassified or appropriated. Carpenter's (2008) work on the politics of "wasteland" further shows how state and market actors erase women's ecological knowledge by labelling fertile or communally used spaces as "empty," thereby legitimising their conversion. The loss of land is never only a material or economic displacement; it is a gendered rupture that severs the intimate ties through which women sustain households, livestock, the commons, and the relations that bind these worlds together. As women are more directly connected with natural resources especially in rural world other than forest, women's role in managing and conserving environment is undeniable (Mago and Gunwal 2019). The other aspects of women in literature were discussed in terms of landownership (Agarwal et al. 2021) and women empowerment in agricultural terrains where farming is the primary profession (Agarwal 2020). However, the women in villages of Punjab centring Lahore city (as it is the area where research was conducted) depicted a picture of a woman who was confident, articulate and well-aware of on-going social issues more so, through their day-to-day experiences. These women broke the stereotype of a woman which has been created over the long haul, although it merits to be researched more deeply. However, what I have experienced was a breath of fresh air.

One of the interesting aspects which was unwrapped during conversations session and seeing them personally was the equation they shared with men at their houses and their role in decision-making, especially in budgeting monthly expanses, doing savings for rainy days, children's marriages and education. Not even once, women talked about their counterparts with animosity but expressed their emotions akin to marital life, otherwise. They don't present it as a perfect life but acknowledge the hard work of men they put in to provide for their families. They also believe that house can't be a home unless women take care of it. Hence, women called it a partnership and spoke highly of men as their team mates. There are still women from certain castes such as *Jatts* and *Rajput*, where they still don't like their women to work outside the premises of home; however, parents don't hesitate in sending their girls for education despite the fact that education is an expensive task for the majority of families. The other myth associated with women covering their heads as a sign of oppression and dictation from patriarchy was busted. It is women's agency to dress up in way which is modest, and they take pride in it. It is relevant to be inclusive about women's agency for embracing their traditions with pride and own free will (Mahmood 2006), and it needs to be regarded as freedom even if they are out of the scope of Western idea of women's liberation and freedom. The element of '*haya*' is not only in covering their heads but conducting their regular lives as well including animals. The building of separate room for a pregnant cow and taking care of her privacy was personally witnessed. During labour, men and young unmarried girls are not supposed to be there, but elderly women gather to take care of her just like their own daughter. The girls are getting education; nonetheless, doing jobs is still debatable and varies from clans to clans; *Jatts* and *Rajputs* have been against the idea of women working outside the premises of their houses even if it is agriculture. Whereas *Arayi* had a different take. Similarly, now educated girls are working as teachers yet not so rampant. But, the point to ponder is that "women not working to earn money" is deemed as women suppression and a patriarchal move to clench their counterpart from being independent. But there is a very minimum documentation of women who say out and loud that men are the breadwinners, and it is their job to provide, and women already have a lot to do; then, why to add more to their platter. Women working as domestic workers is a rising profession, and wives now have to get out of their homes to provide care and other services to their employer's family rather their own (Farris 2013; Hochschild 2004) due to financial crunch they experienced, especially with the loss of agrarian ecology. It again has raised the question of 'choice'. Are they happily and emotionally depleting their children and families with their absence or are they dragged out of their houses to work for money? The other important issue they brought attention to was about finding suitors for children; especially for daughters, it became even more complicated due to dowry. The child's marriage was not brought before 2018, but in past recent years, every mother talked about it and connected it to greed which has no end. I also came to realise that parent's decision for choosing a suitor for a child is not a gender-based but goes with an approach which says, 'Parent know the best and obedient children trust their elders' decisions', although families do have started asking their daughters, but that 'asking' is more or less 'informing'. Love marriages are still stigmatised which is again gender-neutral. Women are still not included in 'Panchayats' (village councils); however, *Rahmani braderi* in *Durgpura* provided a different picture that again was far from mainstream gender discourse. They mentioned about an elderly woman in the village, who they used to go for the final decision especially after her husband died. Upon enquiring further, they explained the only reason for approaching her for final say was because her wisdom came from her life-long experiences which no younger person could come at par with. The purpose of mentioning these anecdotes in the chapter is to underline that gender discourse, women's experiences, their relations, and their notions of freedom and mobility cannot be understood through a radical feminist lens alone. Nor can any of these experiences be generalised across contexts, let alone from one country to another. One can come across striking differences between women in a village of southern Punjab and Central Punjab; intra-gender experiences show a range going from border to ghettoised settlements and also from a village to upscale locality. Their movement from freedom to confinement is not symbolic but lived, and their bodies register this shift with unmistakable clarity. Postcolonial feminist scholars remind us that emotions are not merely feelings but forms of embodied knowledge, and the women's accounts here reveal how dispossession settles into the body as fatigue, anxiety, restlessness, and grief. The sense of being alone, enclosed, and watched is not a metaphor. It is an everyday reality. The mourning for trees under which they once rested, exchanged stories, found companionship, and released the burdens of the day is equally real. These spaces were not simply physical locations but affective and relational ecologies that held their lives together, illustrating what feminist political ecology identifies as the deep entanglement of gender, environment, and social wellbeing. When they were stripped of land that the state and developers portrayed as a "deal" and a win–win solution, they did not simply lose property. They lost a way of being in the world. What they received in return was a fractured and unfamiliar life, a transformed version of existence they had neither imagined nor been prepared for. This transformation is not only economic. It is affective, spatial, and deeply gendered, revealing how the rupture of dispossession reshapes the very conditions through which women experience freedom, mobility, sociability, and dignity.

The perception of respect and reverence has been redefined that fits well in the capitalist world. Now, 'respectable' is the one who has higher house as a display of wealth instead of a character. *Mewati* have a tradition of not marrying their children to immediate extended families; now, a tradition that had been followed by generations after generations had changed as land transaction by DHA had made it difficult for people to stick to their traditions. Market in land on the periphery has been a known fact, which is why greed for dowry has gone to a next level. Therefore, even *Mewatis* have started doing cousin marriages. One of the debates which usually pointed out that women are not given their due share in inheritance by men is not incorrect totally, but literature does not say a reason why. Till now, women's share in inheritance is dependent upon the owner's economic class and, above all, a person's will. However, one thing that has been constant is the sense of responsibility in men either as a father or a brother towards their sisters and daughters, of providing not just material needs but safety, security and marrying them off and yet continue taking care of them where in need. The gated communities even couldn't ensure safe secure environments which villages used to have. Hence, it is proved that intangibles can be used by real-estate developers for promoting business, but despite demanding hefty cash, residents in gated communities are on their own. This clearly indicates that such intangible assets are the product of a certain ecology that binds society and environment together, and a new cash-based ecology that thrives on individualism has never aimed at producing a culture of collective care but the protection of personal interests.

It is not incorrect to say that land teaches us how to live in a spirit of togetherness without stepping over the fellow dwellers and adjusts with not only humans but also animals and all other forms of life. The architecture of the houses has tremendously changed over time which forced me to think about its effect on human's perception and way of thinking even by being at a same place. It also shows that kind of culture a community wants to have can only happen when material conditions are in accordance with it. For instance, mud houses, which can still be found in the farthest villages, created an environment where inequality was less visible and the pressures to "fit in" through displays of wealth remained minimal. Their simplicity spared people the emotional and financial burden of keeping up with the cemented and painted walls that signal class status in more urbanised settings. In vernacular-architecture terms, these structures were not merely shelters. Scholars such as Paul Oliver (2003) and Amos Rapoport (1969) show that vernacular forms evolve from local ecologies, social norms, and cultural values, and mud houses in Punjab exemplified this: they embodied openness, permeability, and relationality.

The walls of these homes were intentionally low, enabling neighbours to remain aware of one another's circumstances, and they were thin enough for a cry, a quarrel, or a call for help to be heard. Women could speak to each other across these walls, and they could gather within moments when presence was needed, without depending on mobile phones or landlines. This reflects what AbdouMaliq Simone (2004), Ash Amin (2014), and Eric Klinenberg (2018) describe as "social infrastructure": the material environments that support sociability, care, and everyday cooperation. In this context, the architectural form itself sustained social life. The permeability of walls, the proximity of households, and the informality of shared spaces worked together to produce an ecology of mutual support. These vernacular forms were not accidental; they were social infrastructures that enabled attentiveness, reciprocity, and collective wellbeing in ways that contemporary built environments often fail to replicate. (A real connection emerges over time, sustained through hardship and reciprocity, and it rests on a sensitivity to the unsaid, the quiet forms of communication that sustain social life. The open veranda of the houses from where people could see the sky was not just a source of ventilation and natural light to enter, but the beings were also receptive, with accommodating hearts, verandas were enough to keep reminding them to share space with other members, doors are kept open to let people in without hesitation and host is always there to make sure they attend the guests.Open spaces in villages are never difficult to find. Whether under a shaded tree, near a tube well, in the open fields, or during relaxed strolls along unpaved dirt roads, these spaces form part of the everyday social landscape. Outside a small shop or in the porch of a mosque or shrine, such places, although more commonly used by men, function as informal social infrastructure where they gather and spend time together. A hookah still can be taken as a symbol for gatherings. Those open spaces have room for everyone to grow and thrive regardless of age. Women could walk freely without any fear which is a blessing in itself. This fearless environment is not available in gated communities as realtors promised to provide. Children have had play grounds where they would socialise with others, meet elders and spend time in playing with other children in a secure environment where their parents in specific don't have to watch-over as entire village is guarding them. Modern architecture now openly broadcasts the economic standing of its residents. A wooden, weathered door with rusted chains and splintered grains resembles a furrowed brow, each signalling quiet struggle and endurance. By contrast, high concrete walls, compartmentalised rooms, and multiple locks have made households increasingly inaccessible to their neighbours. What appears as progress has instead created enclosure. People find themselves confined within their own homes, reliant on paid help for the most basic tasks and on hired guards to feel secure. Cash, in this new world, works like a double-edged promise. It cultivates the illusion that one can command comfort, labour, and protection at will. Yet it also hollows out relationships, turning them into fleeting and unstable exchanges. Interdependence persists, as it always has, but now without the familiarity or trust that once anchored social life. Money becomes the only thread linking households across class lines, a weak and wavering connection that leaves those insulated behind high walls more anxious than assured. For trust, unlike labour or security, remains an intangible form of wealth that no currency can buy.

The other striking aspect of this development model which allows to draw comparison when it comes to gain and loss account is in terms of intergenerational difference of vision. It has not only distanced one family from another but also created isolated units within each house. Time was another asset which in agrarian setting was available to them as people used to work together, women contributing either through farming or carrying food to fields as a gesture of care, affection and their duty towards a provider. Children were equally involved in activities, such as on field or taking care of an animal which conditioned them to be around elders from whom they could learn skill but life lessons also. Now, teenagers are completely out of the agriculture, and modern education tells them not get their hands dirty. Children at one time wanted to leave the village and to be shifted in better areas, but since villages are no longer a village and they have urbanised, children do not mind living there. Also, cash enables parents to provide electronic gadgets which keep kids included in the modern world but excluded from their intimate relations. Young children have no reason to disdain a life that used to revolve around agrarian landscape; however, the parents and elders of the family seem helpless when it comes to inculcating same values in their present generation with which they themselves were raised. Food preferences have been changed, and day-to-day routines have become urbanised; the lack of space must have triggered it, but mobile phones, computers, video games and Internet are consuming most of their time. Children from pocketed villages were seen playing on roads in DHA phases which was not safer and would not be able to take advantage of roads for long. Raising children as per parents' wish has slipped out of their hands due to multiple factors. The mobile phones and Internet have already put a child in an isolated box where they have their own world which parents can't supervise all the time. Secondly, the modern education which helps producing for the upper classes and completely detaches from their surroundings is almost like planting a seed in artificial conditions far from its roots and a place where it belongs. The words and their concepts in children' minds are different from their elders which is one primary reason for youngsters to have dismissal for the value of community as they can't even comprehend that what does it offer. The community's concern is taken as 'interference', and their care and protective attitude are deemed as 'restrictions' or an anchor to young person's freedom. Villages do have schools for girls and boys, and parents are doing everything in their capacity to provide education to their children. They are already exposed to an environment that could provide them skill and resilience from a community to survive. Now, they find themselves fully dependent on urban jobs where many compete for very few positions, a shift that mirrors what scholars of urban precarity describe as the thinning of livelihoods in expanding cities. Urbanisation produces not only built environments but also new subjectivities marked by individuality and isolation, eroding the collective rhythms that once made everyday life more manageable. This transformation weighs heavily on women. They must navigate the competing demands of paid care work and their own domestic responsibilities, often caring for employers' children while their own receive only what remains of their time and energy. This displacement of care echoes feminist analyses of social reproduction, which show how women's labour is stretched thin across classed and unequal urban worlds.

As in situ migration enlarges settlements and draws in waves of new residents, places that once felt intimate begin to feel unwieldy. The task of watching over one's surroundings becomes more difficult, and families spoke of rising anxieties around security, including the spread of illicit activities such as drug use. Many explained that they left so-called gated communities not because these spaces offered safety, but because they became alien landscapes filled with unfamiliar faces and unpredictable risks. Yet despite these contradictions, the primary motivation behind leaving their ancestral homes remained a hope that their children might grow up in an environment imagined as better suited for the future. In this sense, their movement is shaped by both aspiration and unease, revealing how urbanisation simultaneously promises opportunity and produces profound dislocation.

But perks have their own cost, as villagers are losing their language, a sense of identity and a never-ending struggle to settle in getting conversant to place which is a whole new world, where they are still considered backward and uncivilised. Families who have gained big cash amounts and can afford to have a house amongst the elites have increased their expenditures with added burden to become their 'look-alike'. Combat begins with wearing branded cloths to expensive gadgets. For instance, men who want to expand their connections try to please others by keeping things which are only to show-off to prove that they deserve rich man's time and connect. This is continuous psychological and mental exertion because it requires them to change not only from head to toe but from the core. No matter how nicely they dress up, the inability to talk in language of the posh is a deal-breaker. Eventually, families had to forfeit, and they returned where they belong. Nevertheless, now in present time, decision to have residence 'where' has become more complicated as it's a do or die situation due to increased land prices and property in village. Who are intended to move out is tantamount to burn boats. It is a one-way street where people can't decide on a whim but have to think wisely to be on which side of the road. The periphery has been suburbs for the working class, tenant farmers and peasant, small landholders, a very few big owners but majority landless. The city life and symbols of modernity have never been a conundrum to them as they used to commute to cities for work and they still do. But previously, they were not that exposed, but now they are living in such a way where so-called ideal living remains in front of their houses all the time, and expensive cars passing by keeps reminding them that they are way behind. The beautiful houses, clean streets, and gated entrances just across the road work powerfully on the imaginations of rural dwellers. They cultivate what Appadurai (2004) describes as "the capacity to aspire," drawing people toward visions of a better life that seem almost within reach, as if one could simply stretch out a hand and grasp the moon. This is the seductive pull of desire that Tsing (2024) writes about, a shimmer that promises transformation while obscuring its costs. The invitation is illusory, yet difficult to resist. For those who do cross to "the other side," the lived realities are sharply shaped by gender and generation. Women, in particular, recount feeling trapped rather than liberated, their movements constrained in ways that Massey (1994) reminds us are central to the production of space and power. Daily life becomes circumscribed by dependence on male accompaniment and the need for a car to move even short distances. Instead of expanding their freedoms, the new environment contracts them, producing a quiet loneliness and a deepened sense of enclosure. There is a gamut of experiences akin to gender and generation. Women feel trapped and bound, secluded and lonely, instead of being freer, dependent on men to accompany and a car to commute. This has only given them diseases especially hypertension, diabetes, joints pain and obesity. Since social interaction has diminished to almost nothing, the women they once relied on are no longer present to lend a helping hand, offer a shoulder to lean on, or provide the attentive presence that once held emotional life together. In the language of affect theory, these absences are not merely social gaps but ruptures in the everyday circuits of care and attachment. The subtle energies that flowed through shared gestures, small reassurances, and collective routines have thinned, leaving women without the affective grounding that once sustained their sense of belonging and emotional wellbeing. Women continue to recreate the atmosphere of village life inside their new homes. They place a *charpai* instead of a bed, reach out to familiar community members rather than cultivate new neighbourly ties, and often still prefer to sleep on the rooftop at night, small acts through which they sustain memories and embodied habits of belonging in unfamiliar surroundings.

The study is not a presentation of rural life as a utopia or a fairy-land free of miseries and difficulties. Rural dwellers never have claimed an agrarian life as easy. It is not about seeing cash for the first time since DHA arrives as they were already into selling goods in cities such as milk, cow dung, etc., yet people are lamenting a past life and can clearly see what they have lost in the process regardless of class, gains and professions. Yes, prior to DHA, people did not have options for switching professions and to make money through other means which now they have. Yet, they feel losing everything while they are trapped. Why lives have become more challenging than they have had till 2007? Or is it in the head of urban people that rural life is not only backward and uncivilised but difficult too? The study is the voice of the unheard and faceless people in the eye of planners. The rural dwellers have been experiencing a 'change' that is even hard for them to wrap around their heads, a transformation that has caused upheaval in just less than 20 years. The pace of change that has happened in recent past in these rural settlements due to urbanisation is unfathomable compared to the change happened in 200 years to the Lahore city. This is mind boggling indeed, and inhabitants have got trapped in the process of change. They have shared a history and the perks that agrarian life used to offer for free without being dismissive of farming as arduous and laborious livelihood. Women, while recalling and comparing their former and present life and shifting from mud to concrete houses, never held back from sharing their unconditional relation with their surroundings and expressed that their presence was fundamental in maintaining houses and creating 'home'. Yet, what they share is story of loss—especially a loss of intangibles, which they consider a determinant of quality of life, tied to agrarian mode of production and physical landscape.

They were cash-poor, way rich though. They might not have resources or amenities which make city life 'modern' and hollow, but people in rural realm had more control and independence due to subsistence which earth and rural ecology provided them. They had open lands for animal to feed, they had soil to provide them organic food and large number of mouths to cater in abundance with little or no cash required. Their houses were large enough to keep livestock as their own family member. Now, they have crammed into small places even while knowing that they would not able to get such big houses in cities to live. The food was rich in nutrition, and their drinks were fulfilling and in accordance with seasons instead of modern artificial juices and fizzy drinks. The water in villages was clean and accessible to all for free which now has been replaced by contaminated water and is a source of diseases such as Hepatitis. Now people are eating poison in form of food due to intensive use of pesticides and chemical fertilisers with aim of higher yield neglecting the fact that inhabitants have been deteriorating their health and reducing life expectancies. The villagers knew it well and clearly drew comparison in type of food, its taste, smell and health quotient their food used to offer. The availability of clean water has now become a symbol and measures person's financial strength because water which was fit for drinking was available at 60–90 feet, which has now gone down to 450–550 feet depending on area. Therefore, one needs to have a deep borehole with an electric motor to extract water from the ground. The microclimate has changed and so are the routines. Human beings live by an internal body clock naturally attuned to the rhythms of dusk and dawn. Modern urban life disrupts these circadian cycles, a shift that scholars of political ecology and health recognise as increasing the body's vulnerability to disease. The absence of Covid-19 and dengue in villages, epidemics that gripped the city but did not penetrate rural areas, highlights the protective effects of ecological balance, dispersed settlement patterns, and open-air living. Village residents inhabited what might be called a locally governed socio-ecological world. They collectively managed resources, resolved community issues without external interference, and relied on local production, including food, transport, clothing, and housing, for their everyday subsistence. This self-sufficiency was not only material. It formed the basis of their autonomy, their rhythms of life, and the collective power that once grounded their sense of wellbeing. They were once far from being integrated into larger capitalist growth model which now has incarcerated them, behind unseen bars and invisible cages. Now, they are totally dependent on cash to survive that has happened by depriving them from their land and their freedom to exist independent of market world with added burden. For example, men now have to bring water from places outside the village, people have to buy fodder from the markets that needs transport and fuel, and men and especially women are getting sick as they have no open space to go to take stroll in a fear-free environment. Stress has mounted, and hypertension is becoming common due to competition and need to make and multiply money which again is a never-ending race with no finishing line. One may get to hear that 'greed' has changed people; however, I would like to contemplate over this term 'greed' too. Even if people have become greedy and self-centred, what has changed them? They might have cash influx, but what they get slips away from hands because now they have to spend money even to acquire butter and milk and can no longer get grains and save them till the next harvest or next year; when they were bestowed with plenty, even beggars would not sleep on empty stomachs. Now, they have to be worried for the provision of three meals a day to their children. Hunger is biggest of the evils. Rural ecology had an atmosphere that used to thrive on sharing and reciprocity yet had copiousness of everything. The more they share, the more they gather especially in terms of intangibles. The security that women and children once experienced, and the watchful guidance that kept adolescents on track, emerged from a historically rooted social infrastructure of mutual care and everyday vigilance. As feminist geographers note, such protective ecosystems are produced through dense relational networks, not through walls or surveillance. No gated community or model of urban living can replicate this slow-built ecology of trust, familiarity, and communal responsibility. Villages have a secure environment because no one is stranger to each other; on the other hand, people in gated communities and urbanised living spend money to have safety but are filled with fear to an extent where doors are always locked and no one tried to know a person living next to them. In villages, people are standing shoulder to shoulder through thick and thin; in urbanism, it is survival for the fittest, and every being is on their own. It is rural culture where folks tend to sit together regardless of gender, either under a shaded tree or on the wall of a tube well, on the canal or in open fields, getting together at someone's home or open land in the village; but they do take time out for each other to lend an empathetic ear, to listen to each other which at times is all a person's needs; therefore, there was no sense of loneliness and isolation. The modern living coming from capitalism has taken the most precious asset from people, and that is *time*. It has hemmed in people in such a way where they act and work like machines and have no time to even feel. The isolation due to extreme individualism has put people into depression where they have to spend bucks to be heard to feel lighter. The rural ecology is of use and exchange value, but here I am not taking 'exchange' what profit-hunger world means. The exchange of warmth, trust, friendship, solidarity, respect, and reverence, along with the everyday circulation of goodwill and care, forms the living fabric of village life. These ties sustain the reciprocal relationships between people and the soil, between people and the environment, and among community members themselves. The work of maintaining culture and tradition, and passing these values to the next generation, is central to how a sense of place is created and sustained. Such place-making relies on landscapes that foster connection and interdependence. Once those landscapes are removed or transformed, the ability to maintain these bonds becomes profoundly difficult.

The rural settlements that have experienced change and are under the overwhelming processes of urbanisation have created rifts in environmental metabolism but amongst communities' members too. If not abruptly but gradually, people are drifting apart even if they are still sharing the same space. It is hard to overlook that urbanisation aims at increasing consumerism and attacking means of production, and dispossessing with subsistence is a harbinger to social change like an avalanche. Money enters the picture, and this is where disintegration begins. Social relations are reconfigured as new class formations take shape. Some households rise into newly created layers of hierarchy, while others are reshuffled according to the gains they made through land transactions and how effectively they converted this windfall into social and symbolic capital. As Bourdieu (1984) reminds us, class is not only about income but about the strategic use of resources to claim distinction, and these dynamics of accumulation mirror broader patterns of stratification central to urban political economy. But this is not it, alienation and distance between people also exist because every single household is exposed now. When places change, symbols, meanings and terminologies evolve too. And, capitalist web of relations talks only in the language of numbers (money or cash). This instantly shifts person worth from his character to amount of material accumulation. People instead of refining from inside pay more heed to look presentable so that they can be accepted in the world of the rich. Similarly, to have a beautiful façade of a house has become of paramount importance which automatically puts labels on doors that can be noticed from afar. This further teaches people to choose connections wisely—a further exclusion of people in the hands of their own. It takes toll on psychological and emotional well-being of the dwellers. The ones who are from low-income group start cutting themselves off because they want to keep their self-respect intact. As they begin to feel less wanted and if they go to their rich neighbour's house now, they might misunderstand or be sceptical about their visit. Within a single household, much is unfolding at once. Male members are expected to provide and must leave home for work, yet rising inflation and the escalating cost of living intensify the strain on their earnings. These communities bear a disproportionate burden whenever political transitions and shifting policy regimes reshape economic conditions, making their livelihoods especially vulnerable to the volatility that accompanies each new premiership. In contemporary times, it has become more evident that what villages had was successfully catering its people and meeting their needs by providing means of production and least dependence on market, which is now absolutely the opposite of what they experienced. Hence, women also have to leave to seek work as maids or domestic help to help their better halves to earn more so that they could provide education to their children regardless of gender and could save for rainy days especially for medical emergencies. Hence, mother's care is now getting exchanged for money, leaving her own children unattended. Even village households have started hiring maids, which is a new phenomenon, not yet rampant though.

The exploitation of rural communities has been there even prior to independence, and with time, it has only got worsened. Since partition, there were no land reforms, and none of the measures were ever taken to rectify mistakes by sincerely trying for fair land distribution. Britishers divided agrarian community into set of two, i.e. agricultural and non-agricultural, and very conveniently exempted farm labourers and artisans from the land entitlement or its ownership. Then Green Revolution in Pakistan came in 1965, and several small peasants and farmers and the owners of small landholdings were cut loose and very smoothly accumulated by big landholders who became capitalist landowners. Now in the name of urbanisation, land has been taken away from people, and with the aim of getting rid of settlements, realtors adopt several other tactics without any fear and hesitation. This is a process of creating natural and human monocultures which ultimately serve those who have power and money. Even those who were at advantage and sitting over a valuable land have learnt with time how to make the best out of it and still talk about what have they lost in regard with intangibles and are so convinced that no amount of money can never give them back what they have lost—a sense of identity, recognition, peace, a sense of belonging, relations, warmth, trust, community, safety, a world that works on sharing and not on greed, where people do not have to prove their worth through modem gadgets but through character and their dealings with others. Ecology offered them good food, water and health all free of cost, and now, even health has become a luxury that is only for the ones who can afford it. The stability has been replaced by chaos, peace has swapped place with tension and ecologically stable terrains are now degraded slums where people are seen sitting around trash and then moving out of a village to DHA to depict a picture that is degrading and humiliating for the ones living across the street. It is a constant reminder of acute class difference. Last but not the least, through this study, I would like to reiterate that labelling their experience as 'nostalgia' and rule out the fact that even nostalgia is stemming out of their life-long lives experiences are essential to be taken into account if we want to understand what development project does to communities. Sustainable development goals remain fundamentally flawed because they operate through a business-oriented model that sidelines the epistemologies and lifeworlds of indigenous and local communities. Scholars of decolonial sustainability argue that such frameworks privilege technical solutions over the relational ways of knowing that have long sustained human–environment systems. Political ecology highlights how these models depoliticise environmental protection, reducing it to managerial targets rather than lived practices rooted in place. No resource, landscape, or ecological system can truly be safeguarded without an emotional and cultural attachment to it. As Tuan (1990) and others remind us, these affective bonds cultivate responsibility, care, and stewardship. Such attachments are central to indigenous knowledge traditions, which emphasise reciprocity and custodianship between people and the environments they inhabit. Without recognising and integrating these relational forms of knowledge, sustainability remains shallow and disconnected from the very communities who have historically conserved ecological worlds.

Amidst the fractures of community life, nostalgia surfaced repeatedly in the stories of peri-urban villagers. Yet, this nostalgia was not mere sentimental longing. It was articulated as a mode of critique, a way of holding the present against the memory of more reciprocal and ecologically balanced pasts. Villagers recalled fertile soils, flowing canals, shared labour and kinship obligations not simply to romanticise what had been lost, but to expose the moral poverty of a development model that equates land with cash. Such recollections complicate the narrative of passive incorporation into neoliberal urbanism. They act as a subtle repertoire of resistance—a refusal to accept compensation as adequate or equivalent. In remembering the past, villagers reasserted the value of lifeworlds that cannot be priced: trust, solidarity, belonging and ecological continuity. Nostalgia thus became both a register of grief and a moral weapon, highlighting that the erosion of community life in Lahore's peri-urban frontier is as much cultural and psychic as it is material.

References

Agarwal B (1989) Rural women, poverty and natural resources: sustenance, sustainability and struggle for change. Econ Polit Wkly 24(43):WS46–WS65

Agarwal B (2020) Does group farming empower rural women? Lessons from India's experiments. J Peasant Stud 47(4):841–872. https://doi.org/10.1080/03066150.2019.1628020

Agarwal B, Anthwal P, Mahesh M (2021) How many and which women own land in India? Inter-gender and intra-gender gaps. J Dev Stud 57(11):1807–1829. https://doi.org/10.1080/00220388.2021.1887478

Ahmad S (1970) Social stratification in a Punjabi Village. Contribut Indian Sociol 4(1):105–125. https://doi.org/10.1177/006996677000400107
Akhtar S, Ahmad S, Yousaf H (2019) Role of women in social and agricultural decision-making in the rural areas of the southern Punjab (Pakistan). International Journal of Humanities and Social Science 5:34. https://doi.org/10.14445/23942703/IJHSS-V5I5P106
Akhtar, A. S., & Rashid, A. (2025). Dispossession and the militarised developer state: State and class power on the agrarian-urban frontier of Islamabad, Pakistan *. In The Politics of Urban Land in South Asia (1st ed., pp. 17–38). Routledge India
Alavi, H. (1973). Peasant classes and primordial loyalties. The Journal of Peasant Studies, 1(1), 23–62. https://doi.org/10.1080/03066157308437871
Alias A, Nasir Daud MD (2006) Payment of adequate compensation for land Acquisition in Malaysia. Pac Rim Prop Res J 12(3):326–349. https://doi.org/10.1080/14445921.2006.11104213
Amin, A. (2014). Lively Infrastructure. Theory, Culture & Society, 31(7–8), 137–161. https://doi.org/10.1177/0263276414548490
Appadurai, A. (2004). The capacity to aspire: Culture and the terms of recognition. In V. Rao & M. Walton (Eds.), Culture and public action (pp. 59-84). Stanford University Press
Banerjee-Guha S (2009) Neoliberalising the "urban": new geographies of power and injustice in Indian cities. https://www.epw.in/journal/2009/22/special-articles/neoliberalising-urban-new-geographies-power-and-injustice-indian
Baviskar, A. (2003). For a Cultural Politics of Natural Resources. Economic and Political Weekly, 38(48), 5051–5055
Bernstein, H. (2010). Class dynamics of agrarian change. Fernwood Publ.
Bhattacharya N (2019) The great agrarian conquest: the colonial reshaping of a rural world. Permanent Black
Bourdieu, P. (1984). Distinction: A social critique of the judgement of taste (R. Nice, Trans.). Harvard University Press
Brara, R. (2006). Shifting Landscapes: The Making and Remaking of Village Commons in India. Oxford University Press. https://doi.org/10.1093/acprof:oso/9780195673012.001.0001
Caldeira, T. P. (2017). Peripheral urbanization: Autoconstruction, transversal logics, and politics in cities of the global south. Environment and Planning D: Society and Space, 35(1), 3–20. https://doi.org/10.1177/0263775816658479
Carpenter C (2008) Women and livestock, fodder, and uncultivated land in Pakistan: a summary of role responsibilities. Soc Nat Resour 4(1):65–79. https://doi.org/10.1080/08941929109380743
Chuang J (2015) Urbanization through dispossession: survival and stratification in China's new townships. J Peasant Stud 42(2):275–294. https://doi.org/10.1080/03066150.2014.990446
Comanne D (2021) How patriarchy and capitalism combine to aggravate the oppression of Women. cadtm. https://www.cadtm.org/How-Patriarchy-and-Capitalism-Combine-to-Aggravate-the-Oppression-of-Women
Davidson M, Lees L (2010) New-build gentrification: its histories, trajectories, and critical geographies. Popul Space Place 16(5):395–411. https://doi.org/10.1002/psp.584
Denis E (2018) Urban desires and lust for land. The Commodification of Rural Spaces in the Global South
Devine J, Ojeda D (2017) Violence and dispossession in tourism development: a critical geographical approach. J Sustain Tour 25(5):605–617. https://doi.org/10.1080/09669582.2017.1293401
Doshi S (2015) Rethinking gentrification in India: displacement, dispossession and the spectre of development. In: Lees L, Shin HB, López-Morales E (eds) Global gentrifications: uneven development and displacement. Policy Press. https://doi.org/10.1332/policypress/9781447313472.003.0006
Farahani I (2013) Vanished in gaps, vanquished in rifts: the social ecology of urban spatial change in a working class residential area, Peykan-Shahr, Tehran, Iran. Journal of Political Ecology 20:395–412

Farris SR (2013) Neoliberalism, migrant women, and the commodification of care. Schol Femin Online 11:1–11.2. https://sfonline.barnard.edu/neoliberalism-migrant-women-and-the-commodification-of-care/

Gazdar H (2009) The fourth round, and why they fight on: an essay on the history of land and reform in Pakistan

Ghatak M, Ghosh P (2011). The land acquisition bill a critique and a proposal. 46

Goldman M (2011) Speculative urbanism and the making of the next World City. Int J Urban Reg Res 35(3):555–581. https://doi.org/10.1111/j.1468-2427.2010.01001.x

Grover BR (1963) The Indian economic and social history review: nature of land-rights in Mughal India. Indian Econom Soc History Rev 1(1):1–23. https://doi.org/10.1177/001946466400100101

Gururani S (2020) Cities in a world of villages: agrarian urbanism and the making of India's urbanizing frontiers. Urban Geogr 41(7):971–989. https://doi.org/10.1080/02723638.2019.1670569

Gururani S, Dasgupta R (2018) Frontier urbanism: urbanisation beyond cities in South Asia. Econ Polit Wkly 53:41–45

Hall D (2013) Primitive accumulation, accumulation by dispossession and the global land grab. Third World Q 34(9):1582–1604

Harvey, D. (2003). The New Imperialism. Oxford University Press. https://doi.org/10.1093/oso/9780199264315.001.0001

Hochschild AR (2004) Love and gold. Nouv Quest Feminist. 23

Hull, M. S. (2012). Government of paper: The materiality of bureaucracy in urban Pakistan. University of California press

Hussain A (1984) Land reforms in Pakistan: a reconsideration. Bulletin of Concerned Asian Scholars 16(1):46–52. https://doi.org/10.1080/14672715.1984.10409782

Kennedy L, Sood A (2016) Greenfield development as tabula rasa. Rescaling, Speculation and Governance on India's Urban Frontier

Khan, D. (2022). Political Economy of Expulsionary Urbanization: Subsumption and Estrangement of Spaces in Pakistan. Review of Radical Political Economics, 54(4), 461–478. https://doi.org/10.1177/04866134221093748

Klinenberg, E. (2018). Palaces for the people: How social infrastructure can help fight inequality, polarization, and the decline of civic life. Crown

Kröger M (2012) The expansion of industrial tree plantations and dispossession in Brazil. Dev Chang 43(4):947–973. https://doi.org/10.1111/j.1467-7660.2012.01787.x

Levien M (2013) The Politics of Dispossession: Theorizing India's "Land Wars". Polit Soc 41(3):351–394. https://doi.org/10.1177/0032329213493751

Levien M (2015) Social capital as obstacle to development: brokering land, norms, and Trust in Rural India. World Dev 74:77–92. https://doi.org/10.1016/j.worlddev.2015.04.012

Levien, M. (2018). Dispossession without Development: Land Grabs in Neoliberal India. Oxford University Press

Magdoff F (2013) Twenty-first-century land grabs accumulation by agricultural dispossession

Mago P, Gunwal I (2019) Role of women in environment conservation (SSRN scholarly paper No. 3581051). https://doi.org/10.2139/ssrn.3581051

Mahalingam A, Vyas A (2011) Comparative evaluation of land acquisition and compensation processes across the world. Econ Polit Wkly 46(32):94–102

Mahmood S (2006) Feminist theory, agency, and the Liberatory subject: some reflections on the Islamic revival in Egypt. Etnografica 10:121–158. https://doi.org/10.33356/temenos.4633

Mahmood, S. (2012). Politics of piety: The Islamic revival and the feminist subject (Paperback reissue). Princeton University Press

Malatesta O (2019) "A feminism aimed at liberating all women must be anti-capitalist" [Interview]. https://jacobin.com/2019/10/nancy-fraser-feminism-anti-capitalist-99-percent-majority

Mallik C, Sen S (2017) Land acquisition policy and praxis: the case of Peri-urban Delhi. In: Acharya SS, Sen S, Punia M, Reddy S (eds) Marginalization in globalizing Delhi: issues of land, livelihoods and health. Springer, pp 115–126. https://doi.org/10.1007/978-81-322-3583-5_7

Marglin S (2008) Why thinking like an economist can be harmful to the community. Challenge 51(2):13–26. https://doi.org/10.2753/0577-5132510203
Marglin S (2010) The dismal science: how thinking like an economist undermines community (First Harvard University Press paperback edition). Harvard University Press
Massey, D. (1994). Space, place, and gender. University of Minnesota Press.
Maqsood A (2017) The New Pakistani Middle Class. Harvard University Press. https://doi.org/10.2307/j.ctvgd34n
Mpungose A, Maharaj B (2022) Megaprojects in the context of neoliberalism: socio-economic and spatial impacts of the proposed dig-out port in Durban, South Africa. Urban Forum 33(2):173–195. https://doi.org/10.1007/s12132-021-09442-8
Naràin V (2009) Growing city, shrinking hinterland: land acquisition, transition and conflict in peri-urban Gurgaon, India. Environment and Urbanization 21(2):501–512. https://doi.org/10.1177/0956247809339660
Neef A (2021) Tourism, land grabs and displacement: the darker side of the feel-good industry. Routledge. https://doi.org/10.4324/9780429340727
Neef A, Singer J (2015) Development-induced displacement in Asia: conflicts, risks, and resilience. Dev Pract 25(5):601–611. https://doi.org/10.1080/09614524.2015.1052374
Neef A, Touch S, Chiengthong J (2013) The politics and ethics of land concessions in rural Cambodia. J Agric Environ Ethics 26(6):1085–1103. https://doi.org/10.1007/s10806-013-9446-y
Neef A, Ngin C, Moreda T, Mollett S (2023) Global land and resource grabbing. In: Neef A, Ngin C, Moreda T, Mollett S (eds) Routledge handbook of global land and resource grabbing, 1st edn. Routledge, pp 1–17. https://doi.org/10.4324/9781003080916-1
Oliver, P. (2003). Dwellings: The vernacular house worldwide. Phaidon Press
Petesch P, Badstue L (2020) Gender norms and poverty dynamics in 32 villages of South Asia. Int J Community Well-Being 3(3):289–310. https://doi.org/10.1007/s42413-019-00047-5
Qadeer MA (1996) An assessment of Pakistan's urban policies, 1947-1997. Pak Dev Rev 35(4):443–465
Rapoport, A. (1969). House form and culture. Prentice-Hall
Roy, A. (2009). Why India Cannot Plan Its Cities: Informality, Insurgence and the Idiom of Urbanization. Planning Theory, 8(1), 76–87. https://doi.org/10.1177/1473095208099299
Safransky S (2017) Rethinking land struggle in the Postindustrial City. Antipode 49(4):1079–1100. https://doi.org/10.1111/anti.12225
Sarkar S (2014) Speculation in land and the predicament of capitalist urbanization. SSRN Electron J. https://doi.org/10.2139/ssrn.2462653
Sarkar S (2015) Beyond dispossession. Comp Stud South Asia Afr Middle East 35(3):438–450. https://doi.org/10.1215/1089201X-3426289
Scott, J. C. (1977). The moral economy of the peasant: Rebellion and subsistence in Southeast Asia. Published in 1976. (Nachdr.). Yale Univ. Press
Shin HB (2016) Economic transition and speculative urbanisation in China: gentrification versus dispossession. Urban Stud 53(3):471–489. https://doi.org/10.1177/0042098015597111
Simon D (2008) Urban environments: issues on the Peri-urban fringe. Annu Rev Environ Resour 33(1):167–185. https://doi.org/10.1146/annurev.environ.33.021407.093240
Simone, A. (2004). People as infrastructure: Intersecting fragments in Johannesburg. Public Culture, 16(3), 407–429
Sivaramakrishnan, K. (1995). Colonialism and Forestry in India: Imagining the Past in Present Politics. Comparative Studies in Society and History, 37(1), 3–40. https://doi.org/10.1017/S0010417500019514
Smith N (2010) Uneven development: nature, capital, and the production of space. University of Georgia Press
Steel G, van Noorloos F, Klaufus C (2017) The urban land debate in the global south: new avenues for research. Geoforum 83:133–141. https://doi.org/10.1016/j.geoforum.2017.03.006

Swyngedouw E, Heynen NC (2003) Urban political ecology, justice and the politics of scale. Antipode 35(5):898–918. https://doi.org/10.1111/j.1467-8330.2003.00364.x

Tassadiq F (2022) Producing dispossessed and humanitarian subjects: land acquisition and compensation policies in Lahore, Pakistan. POLAR- Polit Legal Anthropol Rev 45:240. https://doi.org/10.1111/plar.12506

Thompson, E. P. (1971). The moral economy of the english crowd in the eighteenth century. Past and Present, 50(1), 76–136. https://doi.org/10.1093/past/50.1.76

Tsing, A. L. (2024). Friction: An ethnography of global connection.

Tuan, Y.-F. (1990). Topophilia: A study of environmental perceptions, attitudes, and values; with a new preface by the author (Morningside ed). Columbia Univ. Press.

Vij S, Narain V (2016) Land, water & power: the demise of common property resources in periurban Gurgaon, India. Land Use Policy 50:59–66. https://doi.org/10.1016/j.landusepol.2015.08.030

Vijayabaskar M (2010) Saving agricultural labour from agriculture: SEZs and politics of silence in Tamil Nadu. Econ Polit Wkly 45(6):36–43

Zoomers A (2011) Introduction: rushing for land: equitable and sustainable development in Africa, Asia and Latin America. Development 54(1):12–20. https://doi.org/10.1057/dev.2010.97

Chapter 7
Beyond Repair: Cash Cannot Compensate for the Irreplaceable

Contents

Abstract This chapter concludes the book by foregrounding the irreparable losses that compensation cannot cover. Multi-sited ethnographic research in peri-urban Lahore (2014–2022) shows that while villagers received high cash returns for land acquired by elite housing schemes, these payments in the name of 'compensation' concealed deeper displacements that money cannot replace. Fertile soils were stripped, commons erased and agrarian livelihoods dismantled. Grazing animals once held as financial security were forced to be sold, while walls of seclusion and enclosure rose as reminders of who was deemed "not worthy enough" in the neoliberal development agenda. Communities once sustained by reciprocity, kinship and ecological balance now confront alienation, stranded on psychological and emotional islands, experiencing class polarisation and food and health quality decline. Land, far from being an inert asset, is a living ecology and a moral economy. Its conversion into speculative real estate produces visible disruptions—displacement and ecological degradation and invisible losses: fractured kinship, eroded safety nets, fading memory and belonging. The Covid-19 pandemic underscored the precarity of urban life and the resilience of agrarian ecologies. The chapter concludes that development in peri-urban Lahore is less a promise of prosperity than a political project of ruination. What is lost is beyond repair; what is taken is irreplaceable.

H. Javaid, *Land Commodification, Urbanisation and Social Change in Peri-Urban Villages, Lahore*, Advances in Asian Human-Environmental Research, https://doi.org/10.1007/978-3-032-10462-5_7

7.1 What this Study Offers?

This book studied urbanisation and its processes not through a city-centric approach but through the lens of community of villages. This book emphasised that 'urbanisation' by definition and concept has been evolving and is no longer circumscribed to rural–city migration, as with time, process has taken unfathomable pace and shape; hence, it has transcended the bounds of cities and clenched the peripheries and ecologically sensitive zones too unapologetically. This is why, it is important more than ever to give places, geographies and ecologies due recognition as impacts of development project such as urbanisation, gentrification, speculation or land financialisation cannot be generalised. The process of land transaction for urban expansion, infrastructure building, industrialisation and agro-farming has been happening in different parts of South Asia, but we have reported what's new about it in the context of peri-urban villages in Lahore, Pakistan. Its significance lies in the record of the urbanisation process in contemporary context when it has become a 'trend'. The research is a critique of mainstream development under neoliberal policies that focus on profiteering and infrastructure development projects undertaken or endorsed by the State, which plays a major role in further marginalising the marginalised (Tiwary 2024). There is no study from Pakistan which is based on empirical qualitative data to record the process of urbanisation as experienced and manifested as seen through a focus on village settlements. It is a holistic study to document change due to the adaption of neoliberal economic regime by the State policies, and where army as an institution of the State is a major player in the development. However, the focus of the study is not the State as in other studies such as Ahmed (2023), Shahbaz (2022), etc. This study takes a more of an ecological perspective to uncover the material and social change. Gutiérrez (2024) discussed the role of domestic elite in acquiring land from rural areas and using 'financialisation' as an alternative process for accumulation. The focus on a collection of villages in the area helped to map the phenomenon at different stages of transformation and also the differential impacts on particular locales with their own socio-historical context. The 'undeveloped' villages to those totally obliterated by the new gated housing development presented the rural past to the contemporary lived reality of villagers and land. It maps and assesses the losses and gains as experienced by a wide cross-section of village society, including different classes, ethnicities, gender and age. It records the transformation in values, identities, sense of security and attachment to land as expressed by villagers and analyses the same in conjunction with changes in built environment, consumption, routines and demography of impacted communities.

7.2 Places: Lived and Remembered

The rapid transformation of peri-urban villages on the southeast of Lahore provides an opportunity to look closely at the process of change that a 'land' goes through as it urbanises. The land was not productive due to the presence of fertile agricultural soil, but the communities who used to produce helped feed the city. For developers, it was a low-value land that could be developed and marketed for profit. This development however is the harbinger of bigger transformation from where there is no turning back. Land is more than a physical entity and a source of produce and subsistence. For the local communities, land and land-based relations were the very fabric of their lives. Land is not an empty piece of turf; it is an ecology that has its material and symbolic role. It supports a collective life and instils a sense of belonging, responsibility which leads to communal management of resources and ownership beyond the idea of private property. Land is the source of power to community. It cannot be treated as a piece of rug that can be wrapped and put elsewhere (Li 2014). It is highly social in character and pivotal to create an ecology that gives quality of life to villagers.

However, land acquisition is the foremost and primary need for capitalist urban growth as it is for all state 'development' operations such as mining, monoculture farming and infrastructure development. In the market ethos, the developer and state offer cash compensation for land and present it as a 'fair deal'. This so-called compensation externalises not only environmental costs but social costs as well. Harvey says, *'The state has to guarantee, for example, the quality and integrity of money'*, the ultimate goal in neoliberal era, for which State has played role by turning everything into commodity such as land, education, health, etc.

The villages of southeast Lahore were studied as microcosms of socio-ecological change. The research entailed two types of records. The change in land-use, built areas, infrastructure and vegetation was mapped using remote sensing. The data collection on social transformation employed ethnography, the major methodology to record the transformation as experienced and narrated by the villagers. Available local history and publications were also used to develop a picture of the change in the area. Ethnography allowed me to listen to the villagers' tales and change as a consequence of the process of development of the new elite suburbs and the ghettoisation or erasure of villages they once inhabited. Overwhelmingly, for the socially marginalised, it was an account of loss. The capitalist economic model has been fostering human and natural monocultures that are now proliferating under an unchecked model of 'cities as engines of growth'. This process has been extensively researched by critical scholars in geography and environmental studies. A vast body of literature (as mentioned in Chap. 3) is found on displacement, dispossession, cash compensations, the role of state, laws regarding land acquisition and compensation and whatnot! However, a critical gap has been the absence of any record of the experience of the villager's assessment of the socio-ecological outcomes of the development of new housing enclaves of the rich on village lands. The question of 'how' these communities go through change through development which brings cash in their lives is a process that is not well documented. Ultimately, it happens to make people realise that they were independent, and freedom was their strength tied to land—now gone. It then created a necessity for 'modern version' of parameters of respect and value that are totally dependent on material conditions. Further reality hits when they are treated as unwanted entities and sore to aesthetics; therefore, they need to be restricted to maintain the standard. Now, animal market will flourish more as villagers need to sell animals, which were also the insurance to them, as no place and permit for grazing are available. They belong to a different social class and less privileged; now, this bitter fact of life has sunken more when inequality presents itself unabashedly through cash and its ostentatious display. Now, gifts of capitalism have annihilated their space by giving a sense of alienation through gentrification, depression and anxiety and have set in due to competition and never-ending race of making and gathering more and more. This depression has marked its presence in form of high cases of hypertension and diabetes. The loss of open land, commons, and the routine of rural living once supported healthier bodies. Now, this lifestyle is being replaced by rising obesity, joint pains, lethargy, and even emerging cases of cancer. The first cracks in communal life appear quietly - hairline fractures born of rising inequality and sharpening class divides. What once held people together now splinters under the weight of fear: fear of losing identity, of being uprooted, of watching one's world dissolve. Displacement leaves scars that throb long after the move, stripping away the sense of safety that only a familiar community can offer. And in the new, "improved" localities, a different kind of struggle begins, a relentless pressure to reinvent oneself. It demands a shedding of language, of habits, of the rural rhythms that once shaped a lifetime. To belong, one must stay updated, perform modernity, and display enough wealth to earn respect. In this way, money does not mend but instead unsettles; it shakes the very foundations of safety and stability, further polarising an already fractured community into ever-narrower boxes of social class. So many villagers choose instead to drift toward places that at least feel familiar. Yet even there, without relatives or acquaintances, they remain outsiders - caught in a limbo where belonging is always conditional, and the effort to root themselves becomes a journey without a clear shore. The cash nexus and their shift from dependence on land and people to cash and market have made people realise that urban living can provide amenities (which again of poor quality as 'who' gets what is pre-decided), but rural living can offer a good quality living which urbanism can never provide.

7.3 Last Words

All ecologies are subject to change and evolve with time. Villages were the vessels of sustained and stable ecologies built around immobile agrarian production. Rural dwellers and nature were not in competition but in a mutual and harmonious relation. Mother earth took care of its people, and villagers looked after the surroundings which sustained them. The symbiosis provided them quality abundant food, water, health, space, culture, community, security and lesser dependence on a market-based competitive world. The list of perks entwined with agrarian production is long. This case study of villages undergoing transformation studied people in their environments and listened closely to their views. Out of this record and analysis has emerged a damning indictment of recent urban 'development' in Lahore city. The findings underscored the entangled nature of the social and material realms. It also demonstrated the effectiveness of ethnography to access complex social processes which may be unfolding at different stages within a limited geographical terrain. The research helps bring into focus the relationship of the city and peri-urban settlements which are centres of agricultural production. It made me appreciate the need for deeper, broader study of historical village ecologies which have evolved and continue to be defined by the nature of agrarian production and allied livelihoods such as of livestock keeping even when the majority of the residents may not be engaged in these economic activities. These are the spaces now being rapidly erased by urban sprawl in cities of the Global South.

People have not only lost their land; they have effectively surrendered their freedom to a political economy that privileges capital over community. What once served as a space to grow food for their families and sustain their animals has been absorbed into an urban expansion that benefits developers, investors, and state-aligned elites. Access to fresh milk, home-grown produce, and nutrient-rich diets - long foundational to rural health, has diminished. With groundwater contamination and declining water tables, even clean drinking water now demands financial investment, revealing how public resources are being replaced by commodified substitutes in the name of "development." As material conditions shift, so do social relations. Bonds once anchored in mutual dependence and shared labour are fraying under pressures engineered by widening class inequality and land speculation. Respect, once rooted in kinship and communal standing, is increasingly indexed to economic worth - a political recalibration of value that mirrors broader patterns of exclusion. Traditional, share-based agrarian livelihoods that offered autonomy and dignity are being displaced by wage labour and precarious, informal urban work shaped by uneven power structures. These new occupations generate income but deepen insecurity, binding people to an economy in which their vulnerability is not incidental but instrumental. In this transformation, dispossession is not simply a social or economic rupture; it is a profoundly political act. The loss of land destabilises not only material life but dismantles the very social fabric through which identity, dignity, and belonging were once collectively negotiated - remaking citizens into subjects of an order that profits from their displacement.

The arrival of new neighbours-migrant labourers, speculative investors, and affluent homeowners in elite housing schemes, has unsettled the social landscape, fracturing the everyday security women once relied on. Increased traffic, unfamiliar faces, and the loss of community oversight have eroded the subtle, gendered protections embedded in rural life. Even families who received substantial cash compensation and managed to buy homes in these upscale suburbs have not escaped insecurity - least of all women. The influx of money may have improved material comfort, but it has undeniably stripped away the ease and safety that came from being anchored in a familiar moral community. Women now navigate a form of modern captivity: confined indoors by fear, dependent on chaperones or guards, and constantly negotiating a landscape that feels alien rather than home. Inside these new houses - symbols of upward mobility; they report loneliness and rising ailments such as hypertension, diabetes, obesity, and joint pain. Their memories of freedom remain vivid: roaming village fields without fear, sitting under trees, gathering near tube wells or irrigation channels, and catching up with neighbours in cool, communal spaces that belonged to everyone. The political shift is unmistakable. A society once shaped by a moral economy, where many tangible and intangible assets were shared rather than commodified - has been undone by the sudden flush of cash and the total monetisation of land. Villagers have not just exchanged land for money; they have traded away a coherent way of life. In the process, they have been absorbed into an urban order where value is dictated by markets, security is privatised, and the communal protections that once guaranteed dignity and belonging have been replaced by the isolating logic of individual wealth.

The Covid-19 pandemic exposed the deep fragilities of urban life, revealing with stark clarity the resilience of rural ecosystems and the quiet strength of community bonds. In the villages studied for this book, not a single case of the disease was reported. Though not formally investigated, villagers attributed their safety to low population density, shared labour, access to fresh food, and the social support and warmth of relations that, in their view, act as immunity boosters inside-out and define their way of living. These peri-urban settlements around Lahore compel us to reconsider the socio-ecological wisdom embedded in land-based production and the everyday rhythms sustained by it.

Yet these villages stand vulnerable before the advancing force of large-scale capital, sanctioned, engineered, and accelerated by a state committed to neoliberal urbanisation. What arrives under the banner of "development" often imposes the hard discipline of the cash nexus, unravelling the social fabric and ecological equilibriums that have endured for generations. The evidence from this study is unambiguous: monetary compensation does not simply compensate; it destabilises. Its effects exceed what the eye can see; they accumulate silently, while injustices scream to ears willing to hear them, a form of slow, penetrative violence that corrodes lives in order to lubricate an economy designed for the rich. It destroys communal worlds whose very existence stands as an anti-thesis to capital accumulation, erodes reciprocal relations, and degrades the natural systems that once sustained these historical settlements. Nature and society are split apart, administered as separate domains, to serve a global development agenda that thrives on expropriating, cheapening, and extracting value from social, human, and ecological life for the benefit of elite national classes and global markets.

These transformations raise urgent questions about the future of the metropolitan region and its society. As we confront the consequences of relentless and uneven urbanisation, it becomes essential to heed the voices of those who bear its heaviest burdens. Their experiences reveal a truth often obscured in policy discourse: the divide between rural and urban is not an abstract classification but a lived condition. It shapes the daily struggles of millions navigating the disruptions that development brings. In their stories lie not only warnings, but possibilities, reminders of alternative ways of organising life, community, and economy that are rapidly being pushed to the margins. To listen to these voices is not simply an ethical obligation; it is a political imperative for imagining futures that are more just, more humane, and more ecologically grounded.

References

Ahmed A (2023) The rise of military capital in Pakistan: military neoliberalism, authoritarianism and urbanization. Geoforum 146:103846. https://doi.org/10.1016/j.geoforum.2023.103846

Gutiérrez J (2024) Real estate oligarchs: elites and the urbanization of the land question in El Salvador. J Peasant Stud 51(2):489–511. https://doi.org/10.1080/03066150.2023.2252758

Li TM (2014) What is land? Assembling a resource for global investment. Trans Inst Br Geogr 39(4):589–602. https://doi.org/10.1111/tran.12065

Shahbaz S (2022) Deconstructing DHA Lahore: analysing post-1980s military operated housing through three spatial-ideological systems [master thesis, University of Waterloo]. https://uwspace.uwaterloo.ca/handle/10012/18123

Tiwary R (2024) Political ecology perspectives on improving access to environmental commons for the marginalised | economic and political weekly. Econ Polit Weekly 59:17. https://www.epw.in/journal/special-articles/political-ecology-perspectives-improving-access.html

Appendices

Appendix A[1]

[1] Land Type: Military/Govt./Stamped/*Sarkaari* = GL; Ownership/Private/Registered/*Maalki* = PL
Status: Village = V
Walled off/Ghettoised/ Pocketed = P
Erased/Removed = E
Soon-to-be erased/removed = SE
Gated Communities (GC); Urbanised locality (UL)
Communities: Christians = C; Muslims = M
Facts: Villages before partition = Old; Post Partition = New
Amenities: Electricity available = Yes = E (Y); Electricity unavailable = No = E (N); Gas available = Yes = G (Y); Gas unavailable = No = G (N)
Border villages have no canal waster, and these village use tube wells for irrigation. Similarly, Shamlats are only present in privately owned area but not on governmental lands.

H. Javaid, *Land Commodification, Urbanisation and Social Change in Peri-Urban Villages, Lahore*, Advances in Asian Human-Environmental Research, https://doi.org/10.1007/978-3-032-10462-5

Place	Land type	Status	Communities	Land Taken	Facts
Burj	GL	V	Arayi	—	Since 1912; 250–300 houses
Chughalpura	GL	V	Mehar, Sakkay, Rajput, Christians, Arayi	Land has been up for sale illegally	Old; name changed to Habibabad in 2000; 50+ houses; electricity came in the late 1990s, no gas, no government school. Mela at Baba Shah Noor's at Ashiyana Qaid in Bhadun
Jindri	GL	V	—	—	Old; Noor Shah Wali's shrine, annual *Urs*
Rakh Padri	GL	V	—	—	Waste pit (LSE used to dump there, it is no longer there)
Padhana	GL	V	Mostly Malik Awaan and Jatt; also Arayi, Meo, Christians	—	Old, Border Village
Pathanke	GL	V	All Mewati	—	Old, Border Village
Sarja Marja	GL	V	Arayi, Jatt, Malik—many migrated to Durgpura		Old, Border Village
Jahman	Empty to PO	V	Rajput Bhatti, Meo; artisanal castes are here in majority such as ironsmith, weaver, carpenter, lumberman, etc. however, four major castes are Arayi, Taili (involved in oil pressing), Rahmani and Mughal		Old, border village, with the population around 12,000 People migrated from the other side of the border and inhabited the place as majority. They were into livestock and agriculture. When people came here, none if it belonged to the state. Later, they made a system called 'Joray ki Zameen' (land for the pair), according to which 15 acres was allotted to one pair. The leftover land was allotted to retired army officials. During Ayub's regime in 1958, he gave border land to army personnel; later, they hired contractual farmers to cultivate land Famous for red chilli production

Pathanwala	PO	E	Meo, Rajput, Pathan	2013	DHA dealt directly as there were 10 houses; Baba Jand peer shrine with graveyard
Heera Singh	PO	E	Meo & Gakkhar	—	Infamous for selling graves; shrunken graveyard is still present Famous for producing high-quality carrot
Jalal abad	—	E	—	1996	Graveyard sold
Sangatpura	—	E	—	—	Graveyard sold
Gobindpura	—	E	Arayi	—	
Haveliyan	PO	E	Mostly Arayi		
Chak Dhira	PO	SE	Jatt, Lohar, Bhatti, Faqeer		Old, DHA provided electricity but no gas, wastewater pond
Jhallaran	PO	SE	C & M	2002	Settlement area is still in squabble with DHA
Jindra Kalan	GL (primarily)	Some of it is taken over by DHA	Arayi		It is walled off from one end but has the entrance to the village from within DHA phase, rest of it is connected to Jindri
Natha Singh (Fateh Jang Singh)	PO	SE, one house standing	Majority Punjabi, a couple of Mewati houses, but Meo and then Jatt were major landholders; others were Saqqay, Jolahay, Taili, Malik	2006–2007	Old, the last Meo family had taken stay over land for the past 10 years, waste water pond still under family's possession. Graves were sold (especially by the poor communities) and shifted
Chak Bhart	PO	P	Rajput (Khokhar)	2003, built-up in 2007–2008	Old
Lehna Singh	PO	P	Meo, few Punjabi and majority Arayi, 10 Christian households	2001–2002, by 2006 all agricultural land was sold	Old; 200 houses before DHA came and migration started. More than half of the households are in real estate. Main govt. routes were sold too. Legal case due to electricity
Dahoorwala	PO	P	Rajput	2009/10	Old; brick kilns active

(continued)

Place	Land type	Status	Communities	Land Taken	Facts
Kohriyan	PO	DHA has not entered yet, however, people themselves are buying land there			Mobile tower
Harpalke	PO	P	Rajput, Meo, Malik and others such as cobblers, sweepers (changar), ironsmith	2004–2005	Old; it has a shrine where annual festival happens on 22nd of *Haar* (eighth June–eighth July). They have electricity but no gas as DHA did not let that happen
Kalasmari	PO	P	Rajput, Jatt	2002	Old; a part of the graveyard is buried under Green City mall. People fought to save whatever they could
Bhangali	GL	P	Shia & Sunni Qureshi, Awaan,	2007–2008	Old; Mobile tower
Chachuwali	PO	P	Mewati; Rajput Meo	2006/7	New; gas, electricity available. It will be removed in future as masterplan shows its name only written on road grid
Tibba	PO	P	—	—	
Mananwala	PO	Urbanised locality (UL)	—	—	Arable land was taken over by DHA; urban settlement
Roranwala	PO	P	—	—	A part of it was consumed to develop Shuhada town. The settlement surrounded by Askari, DHA 9
Liddharh	PO	P	Rajput, Arayi, Mewati	2002	Old; big settlement
Gohawa	PO	Settlement	C & M	2006	1000s houses in 15 years, located on Bedian road
Mandianwala	PO	Settlement	Majority of Mehar Arayi, a couple of butt and potters' households		16 k houses; now all landless. However, there were no big landholders, neither any migrants from the outside. Most of the land was bought by the paragon (gated community) Waste pit was in the village, now no longer operational. Famous for tobacco and potato production

Haer	PO	Hub for farm houses of the elite	—	Since 2000–2001, market in land took pace	Most of the farm houses of known beings are also in Haer
Mota Singh	PO	V	—	—	Old; lies close to DHA phase 9—Shuhada town Known for selling graves
Durgpura	PO	Walled off from one side + on the main Barki road	C & M; Potter, Sakkay, Malik, Ghauri, Bhatti, Rehmani, gill (Jatt gill)	2006	Old; disputed Christian's community graveyard
Charrar	PO	P-SLUM	Malik, Rahmani, Christians	1977	Old and the first village from where acquisition began. Also known for abundant chickpea production
Saduwala & Ghawindi (at one time, these both villages were twins, later villages separated)	PO	P; Saduwala is the settlement only, whereas Ghawindi is still a proper village on Barki road, closer to the border	Jatt, Rajput, Saqqay Entire area of a village (51 ha = 5 Murabba) was owned by a Pathan who lived in Lahore, he distributed land to people on contract for cultivation. Later, he sold agricultural land to DHA	2012	Old; Sikh dominated; its former name was Ghawindi Khush-haal Singh. Annual festival on 31st may *(Bisakh-Jeth)*; however, open land is no longer available; therefore, festival is shifted to another village, Ghawandi
Kamahan	PO	P	—	—	It is a settlement of the lower-income group. It is known for scandalous dispossession without compensations
Loharpura	PO	E	—	—	E

(continued)

Place	Land type	Status	Communities	Land Taken	Facts
Hadiara	Mixed; PO predominantly and the land towards border is GL.	V	—	—	Old; largest village in Lahore district with the population of 40 K in 2016 (which must have increased by now)
Phuller wan	PO	Settlement is left, but the arable land has been completely sold out			It is situated on Barki road, surrounded by imperial garden, paragon and DHA phase 8 It also had a waste dump which is no longer functional
Haveliyan-Karbath	PO	V	Mostly Mewati		Old; nearby Dera Chahal
Guldasht town	PO	GC	—	—	Upscale gated community; locality for the middle class
Paragon	PO	GC	—	—	Upscale living, people from villages also migrated to paragon who have better sources of income or children are all educated and into jobs
New Abadi Barki (NAB)	PO	GC	—	—	Low-income group, usually wage labourers or people with very small landholdings, sold land to DHA, built houses in this new area
Punjab co-operative housing society	PO	GC	Mix of Jatt and Meo and plots bought by army persons also in the area		For the middle class. People migrated from villages such as Lehna Singh to this locality; with better businesses and once had big landholdings
Barki	PO	Settlement. Agricultural land has been sold	Mostly Rahmani (potter-caste)	Around 2011–2012	In the year 2021, a fight happened between young lads of the village who were protecting the main entrance and the investors Barki's population was around 2400–2500, which spiked to 5000 within one-and-a-half year (in 2016) as people from Sangatpura and Haveliyan shifted to Barki

Appendix B

Table to enlist interviews over the period.

Interviewees	Location	Number of participants	Date	Codes
FWO employee	Roranwala	1	2014	FWO/E
DHA employee	Telephonic	1	2014	DHA/E
Lawyer + DHA investor	Telephonic	1	01-10-2021	
Big landlord	Natha Singh	1	14-04-2021	
Wife of a wage labourer		1		W-Natha Singh
Service staff at uni.	Saduwala	1	08-03-2021	
	Kohriyan	1		
	Gohawa	1	09-03-2021	
Old woman	Heera Singh	1	28-09-2016	
26 yrs. man	Chachuwali + tele.	1	2021 onwards	
Small shop owner	Durgpura stop	1	02-04-2021	
Old teacher	Chachuwali	1	05-04-2021	
Old man	Dera Chahal	1	2016	
Lawyer	Durgpura	1	2016, 2021 onwards	L-Durgpura
Broker	Pathanwala*	1	2014	
Rental shop owner	Written excerpt	1	2014	
Hen slaughter station	Dera Chahal ***	1	2014	
Hairdresser	Charrar	1	25-09-2016	
		1	2021	
Farmer	Burj	1	04-07-2022	
Old resident	Durgpura	1	20-06-2016	
Home-maker	New Abadi Barki (NAB)	1	09-05-2016	
Businessman	Barki	1	19-07-2016	
Big landholder cum realtor	Lehna Singh	1	21-07-2016	
		1	12-08-2018	FM-Lehna Singh
Tenant woman in quarters		1		
Flour shop-owner (Kalasmari)	Barki road	1	2021	
Maid (sheik)	Charrar	1	25-09-2016	
Street-seller (home-made toys)		1		

(continued)

Interviewees	Location	Number of participants	Date	Codes
Blacksmith	Loharpura	1	18-07-2018	
Resident	Jahman	1	26-01-2019	
Cattleman	Bhangali	1	07-04-2021	
Corn-seller woman (originally from Harbanspura)	Barki road	1	31-03-2021	CS
Established greengrocer/meat seller		1		
Potter (Bauwala)		1		
Works in a nursery		1		
Construction supervisor	Chachuwali	1	27-05-2021	
Young teacher		1		
DHA security guard		1		
100-year-old man		1	24-05-2021	
		1	22-05-2021	
Old woman		1		
Old teacher		1	05-04-2021	M-Chachuwali
Wadera	Jhallaran	1	20-03-2021	Wd-Jhallaran
Historian	Khoj Garh	1	25-05-2021	H-Khoj Garh
Man (originally from Kasur)		1		Khoj Garh
Owner of food corner in a village	Jhallaran	1	06-06-2022	Jhallaran, 2022
Old Christian man		1		C-Jhallaran, 2022
Rich agriculturist	Chak Bhart	1	25-07-2022	Chak Bhart, 2022
Jatt	Chak Dhira	1	19-03-2022	Chak Dhira, 2022
Poet	Kalasmari	1	12-06-2022	Poet-Kalasmari, 2022
House help (Mehar town)	Migrated from Mananwala	1	16-04-2022	
Christian woman	Jhallaran*	1	2021	
Resident	Tibba *	1	20-05-2021	
Young man	Chachuwali**	1	Late 2021	
DHA construction site supervisor	DHA (written)	1	2021	DHA/S, DHA- 2021

One-on-one interviews = 56

Focus Group Discussions (F)

Interviewees	Location	Number of participants	Date	Codes
Men	Haer	3	24-07-2016	FM
Women	Bhangali	2	28-06-2022	FW
Men	Revenue office	2	22-06-2021	FM
Women	Chachuwali	3	20-05-2021	FW/1
Mixed	Chachuwali	2	22-05-2021	Fi
Men	Chachuwali	5	15-03-2022	Gujjar-Chachuwali, 2022
Men	Chak Dhira	2	06-06-2022	FM
Women	Dera Chahal	6	14-04-2021	FW
Women	Dahoorwala	10	28-05-2021	FW
Men (Gardner and lawyer)		2		FM
Mixed	Durgpura	3	27-06-2022	Fi/22
Mixed (Rehmani)	Durgpura	10	20-07-2022	Fi/R
Men	Durgpura	10	17-03-2021	FM
Mixed		2		Fi/21
Mixed	Guldasht town	8	21-10-2016	Fi
Women	Haveliyan	4	19-03-2022	FW
Men	Kalasmari	2	12-06-2022	FM
Men	Kohriyan	5	30-04-2016	FM
Women (Kalasmari)	Paragon	3	18-07-2022	FW
Men	Patwarkhana	5	23-11-2016	FM
Mixed (host's family)	Sarja Marja	7	23-03-2021	Fi*1-Sarja Marja
Mixed (haveli)		4		Fi*2
Women (neighbours)		5		FW*N- Sarja Marja
Women		3		FW-Sarja Marja
Mixed	Pathanwala	3	21-07-2018	Fi
Mixed	Chachuwali	2	19-05-2021	Fi
Men (rental place as restaurant)	Barki road	2		FM
Men	Dahoorwala	2	28-05-2021	FM
Men	Kalasmari	5	18-07-2018	FM
Mixed	Habibabad	10	04-07-2022	Fi
Women	Durgpura	10	20-06-2016	FW
Mixed	New Abadi Barki (NAB)	4	09-05-2016	Fi
Men	Rakh Padri	8	04-05-2016	FM
Women		4		FW

(continued)

Interviewees	Location	Number of participants	Date	Codes
Mixed	Kohriyan	5	30-04-2016	Fi
Women		4		FW
Men (soil group)		5		FM
Women	Dahoorwala	5	2018	FW
Couple	Rental fruit stall, Charrar	2	25-09-2016	Fi
Women (landholders, farming)	Heera Singh	3	28-09-2016	FW
Men	Mandianwala	5	05-10-2016	FM
Men	Harpalke	5	30-04-2018	FM
Mixed	Burj	4	01-08-2018	Fi
Women (Gohawa, Durgpura)	LSE	2	10-03-2021	FW-LSE
Men (LSE drivers; Hadiara, Bhangali, Chak Bhart, Barqa Qalan)		4	March 2021	FM-LSE
Men (farming)	Jindri	5	March 2021	FM-Jindri
Men	LDA'S officials	4	2021	FM-LDA
Men (Sardar, patwari)	Padhana, Pathanke, Sarja Marja	4	27-06-2021	FM-border villages
Men (old Lumberdar and group)	Pathanke	4		FM-Pathanke
Middle aged men	Jindra Kalan	2	09-05-2022	FM-Jindra Kalan
Aged woman		1		W1-Jindra Kalan
Middle aged woman		1		W2-Jindra Kalan
Old men	Jindri	3	16-05-2022	FM-Jindri, 2022
Women	Chachuwali Written notes	8	2021	FW/2
Women (calf's birth)		6	2021	FW/3
Men (DHA-land providers)	Telephonic	2	14-03-2024	FM-Land Providers, 2024

Focus group discussions = 56
Total number of interviews = 112

Codes
Focus group discussion = F; Women = W; Men = M; Mixed = i

Appendix C

Punjabi Months	English Month	Farmer's Seasons
Bisakh	End thirteenth month to 21st April	Summer harvest and planting
Jeth	22nd April–seventh June	
Har	8th June–eighth July	Leisure, marriages
Sawan	9th July–14th august	Monsoon rains
Bhadun	15th august–14th September	Waiting
Asuj	15th September–30th November	Autumn harvest and planting
Katak		
Maghar	1st December–28th February	Winter, marriages
Poh		
Magh		
Phagan	March 1st–31st	Spring planting
Chet	Thirteenth month	

(Eglar & Chowdhry, 2010)

Appendix D

The following satellite images[2] are used as examples for showing villages at different stages of development:

[2] Corona Images from 1965 on the left-hand side and Sentinel-2 on the right-hand side from 2023

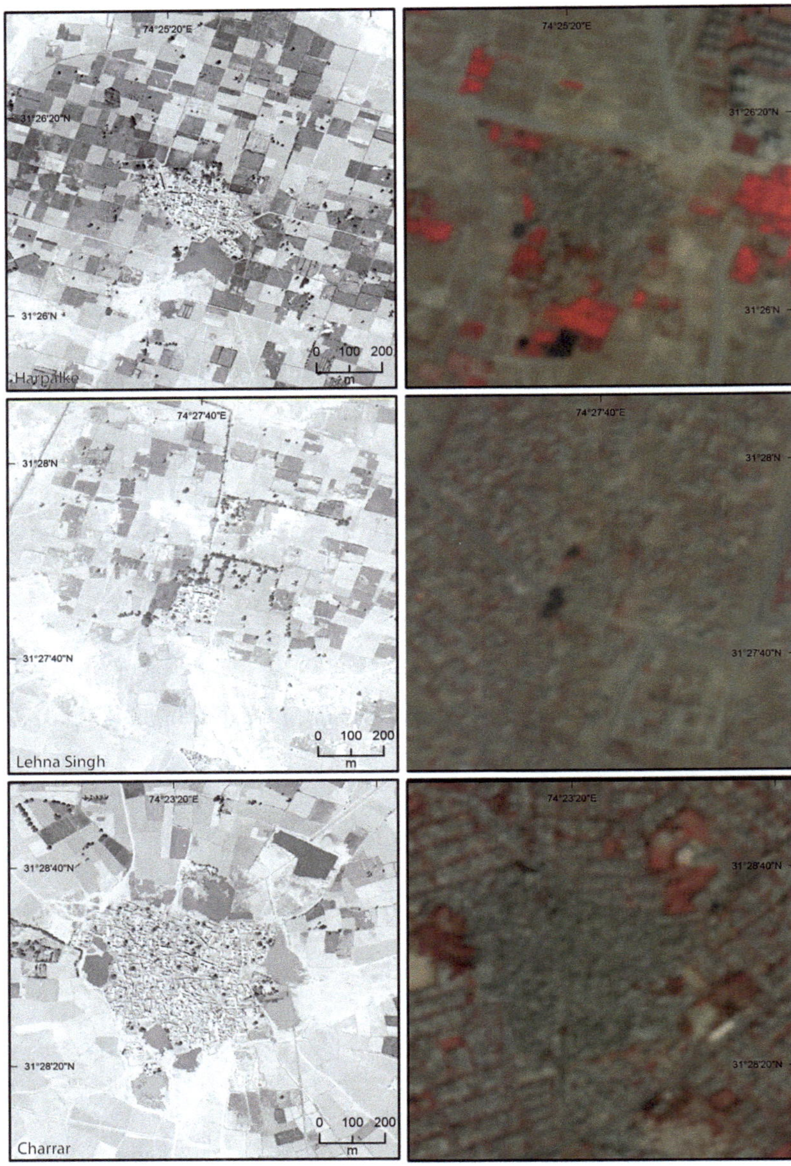

Pocketed: Harpalke (top), Lehna Singh (middle), Charrar (bottom)

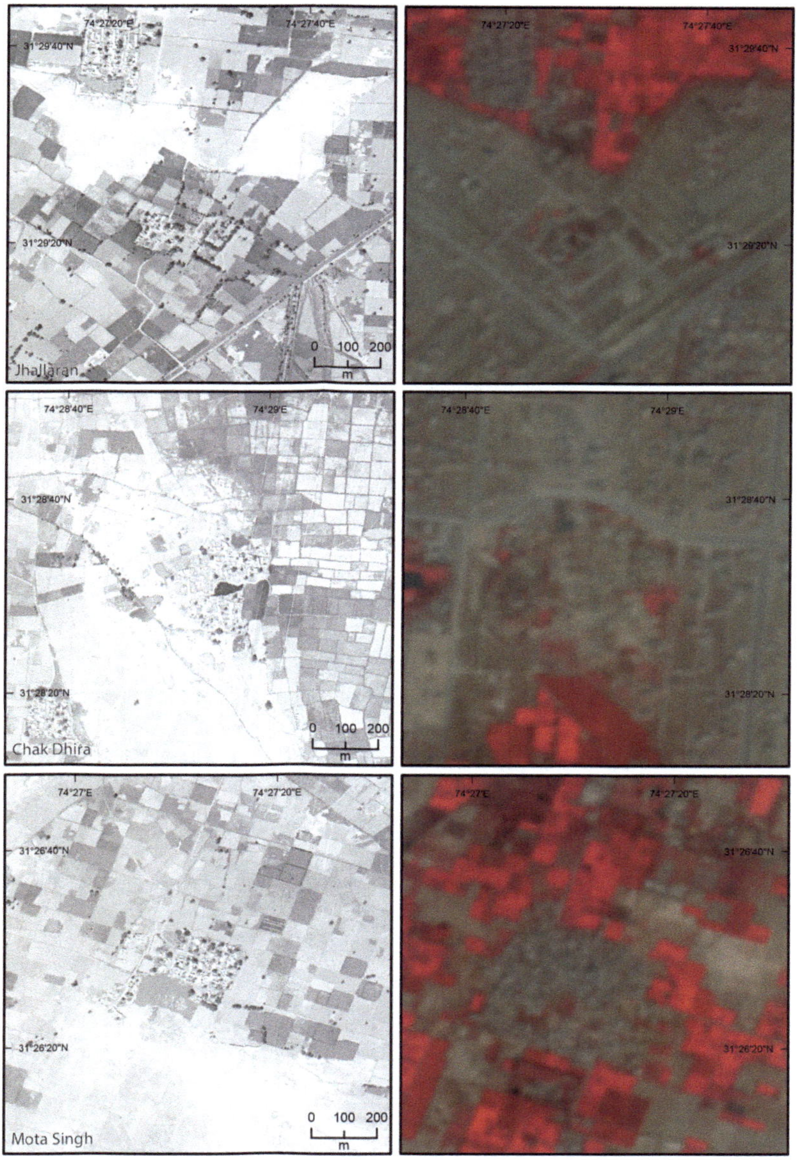

Remnants: Jhallaran (top), Chak Dhira (middle), Mota Singh (bottom) (Mota Singh/Mota Singhwala village is classified as main Mota Singh or Mota Singhwala and Chota Mota Singh. A part of the former village was once taken for the Shuhada Town construction, yet it exists. It is getting sandwiched between DHA sectors, Pakistan Kidney and Liver Institute (PKLI), farm-houses, etc., located on either side of Bedian road.)

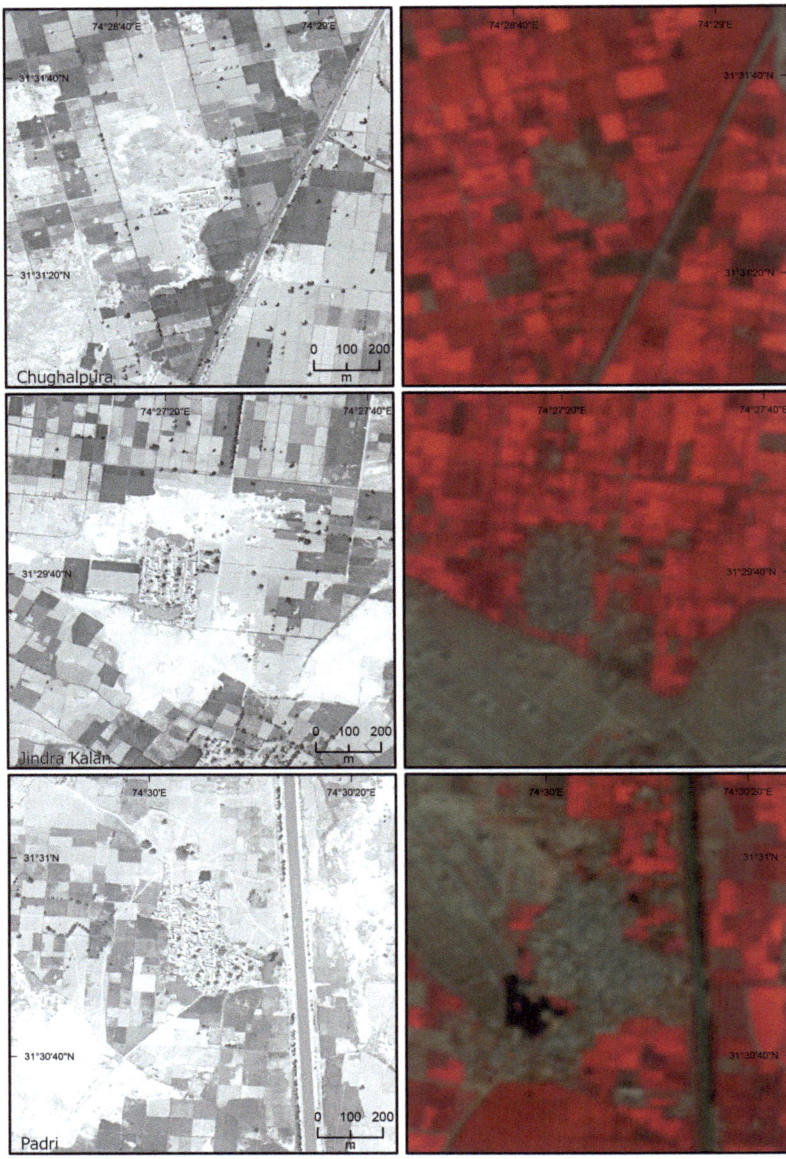

Still rural in character: Chughalpura (top), Jindra kalan (middle), Padri (bottom)

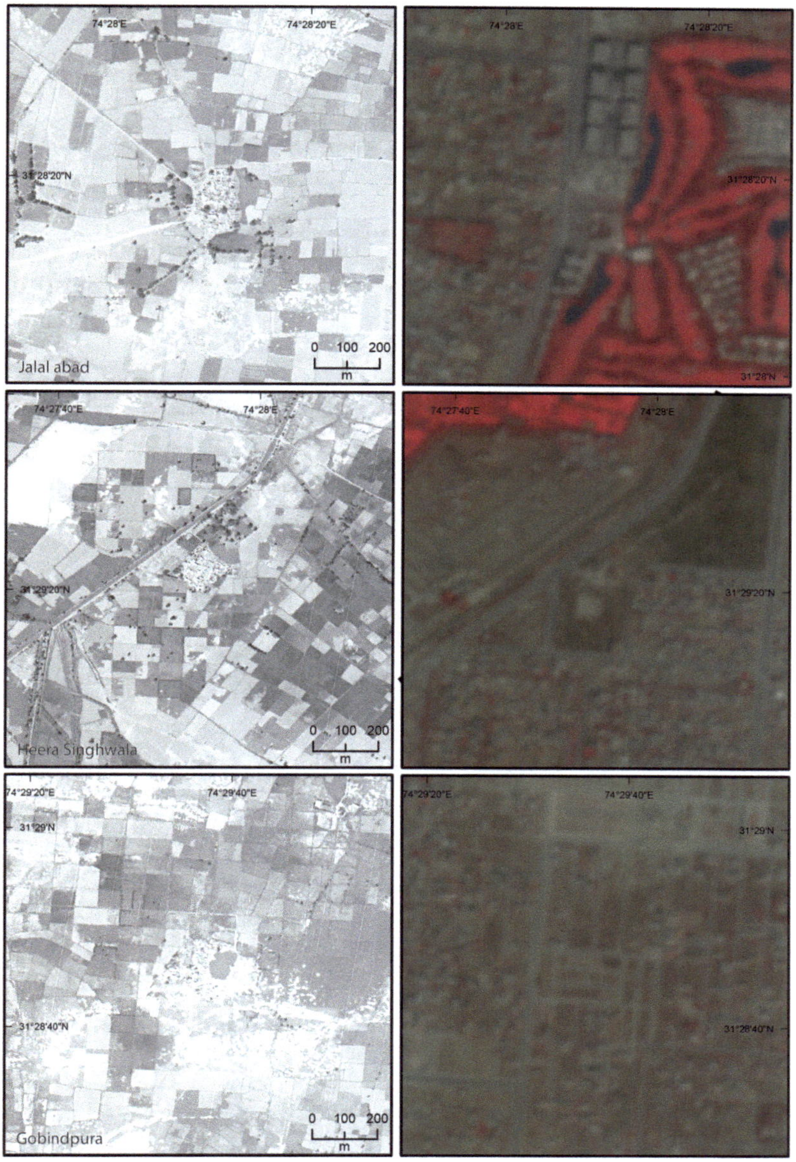

Removed: Jalal abad (top), Heera Singhwala /Heera Singh (middle), Gobindpura (bottom)

MIX
Papier aus verantwortungsvollen Quellen
Paper from responsible sources
FSC® C105338

If you have any concerns about our products,
you can contact us on
ProductSafety@springernature.com

In case Publisher is established outside the EU,
the EU authorized representative is:
Springer Nature Customer Service Center GmbH
Europaplatz 3, 69115 Heidelberg, Germany

Printed by Libri Plureos GmbH
in Hamburg, Germany